FORSTHOFFER'S COMPONENT CONDITION MONITORING

FORSTHOFFER'S COMPONENT CONDITION MONITORING

MICHAEL FORSTHOFFER

ELSEVIER

Butterworth-Heinemann
An imprint of Elsevier

Butterworth-Heinemann is an imprint of Elsevier
The Boulevard, Langford Lane, Kidlington, Oxford OX5 1GB, United Kingdom
50 Hampshire Street, 5th Floor, Cambridge, MA 02139, United States

Notices

Knowledge and best practice in this field are constantly changing. As new research and experience broaden our understanding, changes in research methods, professional practices, or medical treatment may become necessary.

Practitioners and researchers must always rely on their own experience and knowledge in evaluating and using any information, methods, compounds, or experiments described herein. In using such information or methods they should be mindful of their own safety and the safety of others, including parties for whom they have a professional responsibility.

To the fullest extent of the law, neither the Publisher nor the authors, contributors, or editors, assume any liability for any injury and/or damage to persons or property as a matter of products liability, negligence or otherwise, or from any use or operation of any methods, products, instructions, or ideas contained in the material herein.

Library of Congress Cataloging-in-Publication Data
A catalog record for this book is available from the Library of Congress

British Library Cataloguing-in-Publication Data
A catalogue record for this book is available from the British Library

ISBN: 978-0-12-809599-7

For information on all Butterworth-Heinemann publications
visit our website at https://www.elsevier.com/books-and-journals

Publisher: Matthew Dean
Acquisition Editor: Brian Guerin
Editorial Project Manager: Ali Afzal-Khan
Production Project Manager: Sujatha Thirugnana Sambandam
Cover Designer: Miles Hitchen

Typeset by SPi Global, India

Dedication

This work is dedicated to my father, William E. Forsthoffer, who developed the concept of component condition monitoring (CCM) in the early 1990s and uses this method for evaluating machine condition as well as troubleshooting today.

Contents

About the Author

In his 20+ years of experience, Michael has had the opportunity to be involved in design/selection of or field troubleshooting of all of the major types of rotating equipment. He spent 4 years with John Crane, 3 of which consisted of being the on-site seal reliability engineer at Hovensa Refinery, St. Croix, VI. He worked as a rotating equipment specialist for Forsthoffer Associates, Inc. from 2009 to 2015 and has been involved in selection of equipment for new and revamp projects, troubleshooting, and site-specific training worldwide for all types of rotating machinery including pumps, compressors, steam turbines, and gas turbines. He is currently the President of Forsthoffer Associates, Inc., whose goal is to continue to bridge the gap between vendor and user by bringing design and field knowledge to the user in a practical manner to assure maximum rotating equipment reliability and safety.

Preface

The purpose of this handbook is to present the concept of component condition monitoring (CCM) and provide the appropriate parameters to monitor (and preferably trend) to discover a minor issue before it turns into an unplanned shutdown. The concept of CCM, as presented in this book, can be utilized for all types of rotating machinery, making this an all-encompassing tool even when presented with issues on machinery that the reader is unfamiliar with.

Along with information on what and why to monitor, this book also includes spreadsheets with calculations to monitor each component for each type of machine. The spreadsheets are available with embedded calculations to each owner.

Finally, do not think this book is only intended to be read/used by rotating machinery engineers, as operations, maintenance and even I&E personnel can benefit from understanding what parameters need to be monitored and WHY.

Acknowledgment

The concept of component condition monitoring was developed by William Forsthoffer in the early 90s and since has been expanded by Michael Forsthoffer over the last 10 + years in order to encompass all types of rotating machinery.

The first and most important person to acknowledge is my father, William E. Forsthoffer, who has over 50 years of experience and is recognized globally as one of the best minds in the industry. It was his concept of CCM that allowed me to learn the ins and outs of all types of rotating machinery very quickly in a practical manner.

I would also like to acknowledge all of Bill's mentors including Merle Crane, Bob Aimone, Walt Neibel, and Richard Salzmann. These gentlemen were influential presences throughout Bill's career (as well as mine) and indirectly were a part of the development of the CCM concept.

Finally, I want to acknowledge my parents (William and Doris) and my siblings (Brian, Eric, Dara, Jennifer, and Donna) who have always been there for me and continue to be my best friends.

CHAPTER ONE

What Is CCM?

Chapter Outline

Everybody in the industry has heard of condition monitoring for rotating equipment. In fact, every facility (whether upstream, midstream, or downstream) has a group of personnel dedicated specifically to condition monitoring of both general purpose and special purpose equipment. In the majority of cases, the condition monitoring group focuses on looking at bearing vibrations, axial displacement, bearing temperatures, and seal leakage, which are all good parameters to monitor, but are typically symptoms of something else happening in relation to the equipment.

In this book we will focus on and discuss in detail the concept of Component Condition Monitoring or CCM, which was developed by William E. Forsthoffer and has been implemented worldwide since the early 1990s. CCM is valid for all types of rotating machinery and utilizes the parameters that every condition monitoring group looks at, but (if utilized correctly) will also add new key parameters that will aid in diagnosing what the cause of change is and save us from having to shutdown prematurely. It is pure predictive maintenance and if properly implemented will be the basis for turnaround planning.

This chapter will cover the basics of what CCM is and how it can significantly benefit your plant by maximizing on-stream time.

THE BASICS OF CCM

CCM is based on the fact that all types of rotating machinery are comprised of the same five major components, which are the (1) rotor (2) journal bearings (3) thrust bearing (4) seals (5) auxiliary systems. This is the case

Forsthoffer's Component Condition Monitoring
https://doi.org/10.1016/B978-0-12-809599-7.00001-4

whether you are dealing with pumps, compressors, steam turbines, gas turbines, motors, or even gears. The concept is to monitor all of the key parameters that affect each of these components and trend them all on the same time stamp. By doing so, we can easily see when a change occurs in one parameter and be able to see what other changes are occurring over the same time in order to explain what is the root cause. That may sound simple in theory; however we need to have all of the components instrumented such that the key parameters can be monitored and trended. Also, in order to see a change in each component, we need to establish a baseline in order to know what a "Good" value is. These challenges will be discussed in detail along with all of the key parameters to monitor later in this book; however let us first discuss how the five major components are the same for each type of machine.

THE FIVE MAJOR COMPONENTS OF VARIOUS TYPES OF EQUIPMENT

Pumps

See Fig. 1.1 which shows a single-stage overhung centrifugal pump. You will see the numbers 1 through 4 on the figure, indicating four of the five major components in the pump. The rotor (1) consists mainly of the impeller,

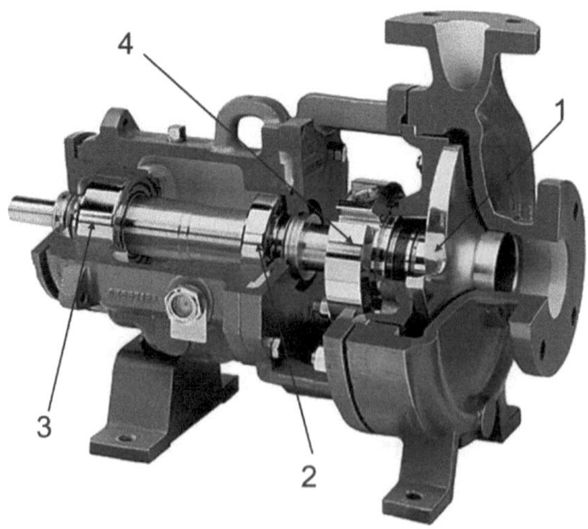

Fig. 1.1 Single-stage overhung centrifugal pump with the five major components highlighted.

which is fixed to the shaft, and provides the energy (head) required by the process. It does this by increasing the velocity of the liquid where it is slowed down significantly and converted to energy in the volute. It meets the required head at a certain flow rate. Therefore, monitoring the flow and head will allow us to tell if the impeller, or rotor, is in good condition or not.

The radial and thrust bearings (items 2 and 3 in Fig. 1.1) are both ball bearings, a type of antifriction bearing. Antifriction bearings are shrink-fitted onto the shaft at the inner diameter and a very tight clearance at the OD into the casing, or bearing bracket. The radial bearing supports the rotor in the radial direction (x–y) and the thrust bearings support the rotor in the axial direction (z). For the case of a single-stage centrifugal pump, the double row angular contact bearing allows for some axial support, so we will call that essentially the thrust bearing. Since the bearings are shrink-fitted onto the shaft, when the shaft bounces, or vibrates, the bearing vibrates and transmits that to the casing. This is why we monitor casing vibration in the x, y, and z plains on pumps fitted with antifriction bearings.

The mechanical seal (item 4 on Fig. 1.1) minimizes leakage of the process to the atmosphere for safety reasons and to keep the liquid being pumped, going to the process. Under normal conditions, this leakage is in vapor form and could be measured in parts per million (ppm). However, visible liquid indicates excessive leakage and should require a seal replacement at the next opportunity.

As you can see in Fig. 1.1, items 2–4 are supporting the rotor (item 1). Therefore, does it not make sense that we look at the rotor on the same time basis of the bearings and seals? Monitoring just the vibration and seal leakage will only tell us that the bearings and seals are not operating properly, but combining that with monitoring performance will give us an indication why we have high vibration or leakage, and prevent future failures.

The major components not shown in Fig. 1.1 are the auxiliary systems (item 5). In the case of a single-stage overhung pump, there are two simple auxiliary systems. Fig. 1.2 shows a typical bearing lubrication system for such a pump. This is a constant level oiler, which automatically distributes oil to the bearings when the level goes below the required level. There is also a sight glass for checking oil level and an oil condition monitoring bottle, which allows us to see if the oil has changed color, and drain out any oil that has accumulated in the bearing bracket. These items should be checked regularly and water drained out if it is visible. If the site glass is not visible, you cannot see what the level is, so you have it cleaned or replaced at next possible opportunity.

Fig. 1.2 Bearing lube system for single-stage overhung pump.

Fig. 1.3 API Plan 11 seal flush system.

Fig. 1.3 shows a typical API Plan 11 seal flush system. This system lubricates and takes heat away from the mechanical seal. If it is not constantly providing cool, clean liquid to the seal, it can overheat and fail prematurely. For this system, there is not much in the way of monitoring, other than making sure temperature is constant through the piping or tubing to indicate flow (a cold spot indicates blockage, usually at the orifice). As we will see later in Chapter 7 of this book, there are many more complicated seal flush systems that need to be monitored regularly to assure they are operating correctly.

So, to sum up the five major components of a pump (and all equipment as we will see next), we can say that components 2–4 support component 1 and component 5 supports components 2–4. It has been the author's experience that over 80% of machinery issues are caused by a problem with component 1 or 5. It has also been the author's experience that components 1 and 5 are the least monitored components of every type of rotating machine. Can you now see the importance of monitoring all of these components together?

Compressors

Now, we will see the beauty of the concept of CCM. Fig. 1.4 shows a typical multistage barrel (or vertically or radially split) centrifugal compressor, with four of the five major components identified.

The rotor (1) consists of many impellers, but just like the pump identified in Fig. 1.1, these impellers produce the energy (head) required by the process

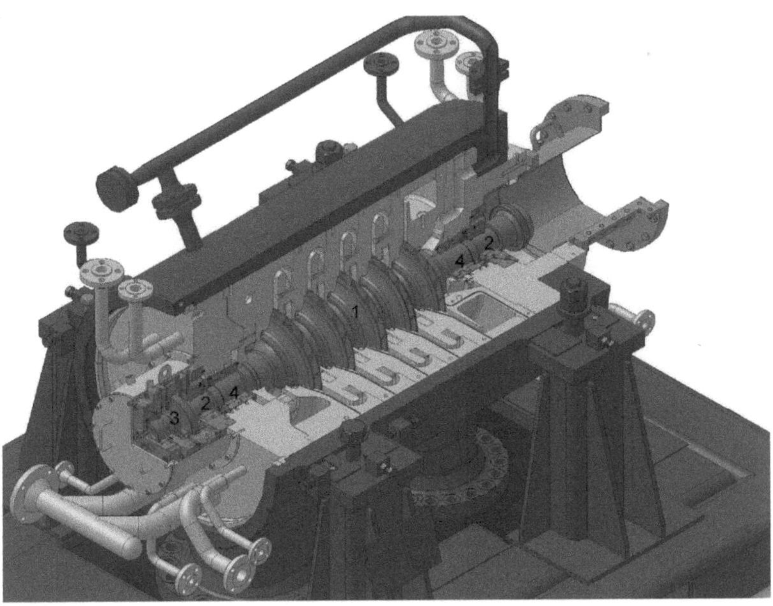

Fig. 1.4 Centrifugal compressor with the major components.

at a specific flow rate. We monitor the same parameters for a compressor rotor that we do for a pump rotor, except we also monitor and trend efficiency as it can easily be calculated in a compressor.

Hydrodynamic bearings are typically used in larger machinery for their heavier load-carrying capacity as shown in Fig. 1.4. The radial bearings, called journal bearings, have a clearance between them and the shaft and an oil film (typically about 20 microns) is what is carrying the load of the rotor. Since the shaft can move in the x-y plane, we monitor the movement of the shaft using radial shaft vibration probes that measure the voltage gap between the probe and the shaft. We also have temperature sensors (RTDs) in the bearings that allow us to monitor how hot the bearing pads are. We will discuss all of the parameters to monitor in Chapter 4, but these bearings support the rotor in the radial direction just like component 2 in the pump.

Component 3 is a hydrodynamic thrust bearing, which supports the rotor and sets the location of the rotor in the axial direction. Just like the journal bearings, the thrust bearing in the compressor shown in Fig. 1.4 uses oil (about 20 micron thick oil film) to support the rotor in an axial direction. We use axial probes to monitor axial shaft position and also RTDs to monitor bearing pad temperature.

Since the mid-1990s, the majority of seals used in compressors are dry gas seals as shown as component 4 in Fig. 1.4. These seals perform the same function as a pump seal, by minimizing the leakage of the process to the atmosphere. We monitor and trend many different parameters to determine the condition of the seals and they will be discussed in more detail in Chapter 6.

The Auxiliary systems for a compressor are typically more complicated than that of a pump. Figs. 1.5 and 1.6 show a lube oil system that provides the cool, clean oil to the bearings at the proper temperature, pressure, and flow rate. This is a very clean example of one; however you can still see how many components it is comprised of that need to be monitored. Many of these components are not instrumented to the DCS, so a lot of local monitoring is required on a regular basis.

Fig. 1.7 shows a schematic of a dry gas seal system that provides clean, dry gas to the seals to prevent unfiltered process gas from escaping to the atmosphere. There are a number of items to monitor in this system, namely the primary vent flow and/or pressure for leakage. This is typically in the DCS and can be trended. We will discuss all of the major parameters to monitor in Chapter 6 of this book.

Fig. 1.5 Pump side view of lube oil system.

Fig. 1.6 Filter end view of lube oil system.

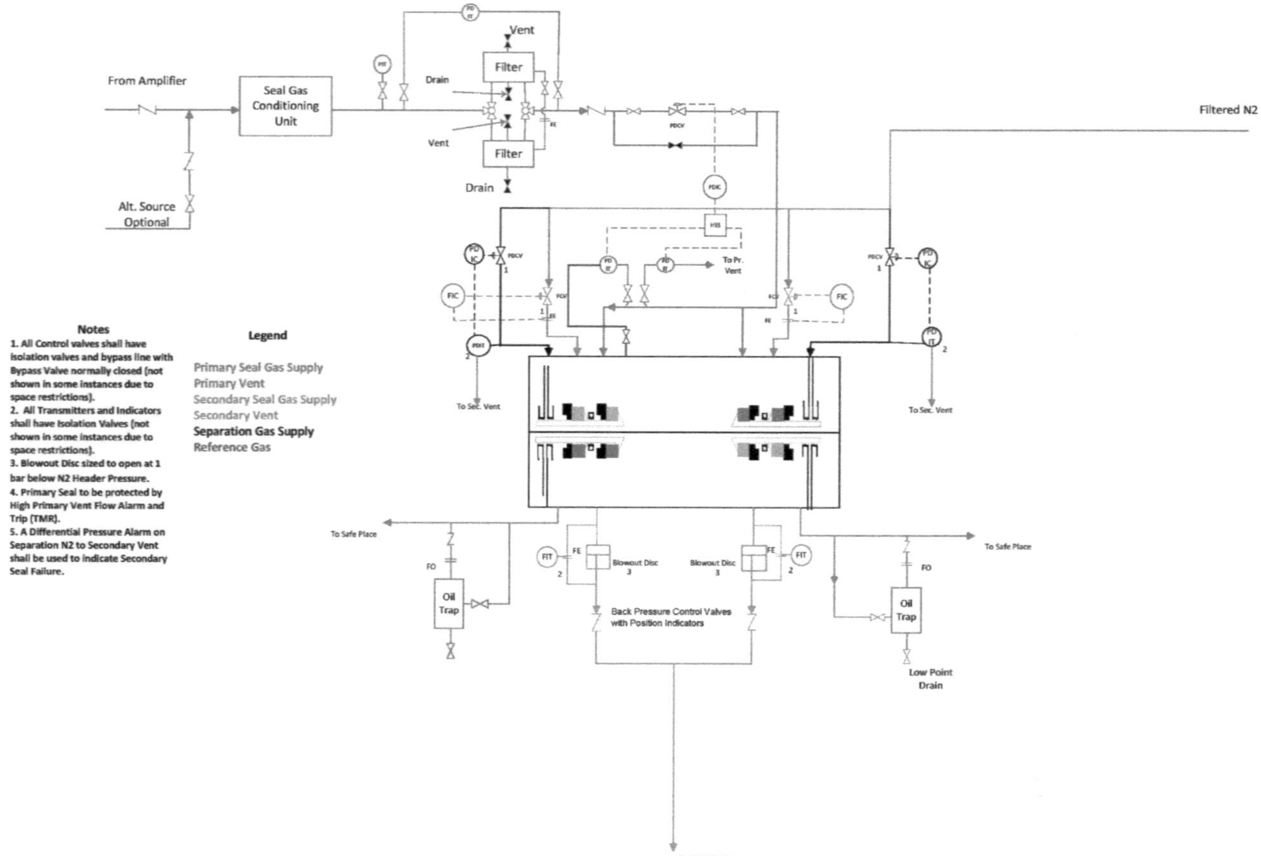

Fig. 1.7 Dry gas seal system schematic.

Steam Turbines

Fig. 1.8 shows an extraction-condensing steam turbine with four of the five major components indicated. We have already discussed these components, since the bearings (2 and 3) are the same and the labyrinth seals (4) are the same as seen in compressors. The rotor (1) is essentially the same as well. Instead of adding energy to the liquid or gas, we are taking the energy out of the steam in this case. We will monitor efficiency as the major parameter for the rotor just like a compressor.

For the auxiliary systems, once again they are very similar to the other machines. You use the same lube oil system as discussed earlier to provide oil to the bearings and the control and protection system for the steam turbine.

The one additional auxiliary system for a steam turbine is the steam seal system that creates a vacuum to suck down the steam leakage to a gland condenser to keep steam away from the bearings (see Fig. 1.9 for a typical steam seal system schematic). The key parameter on this system to monitor is the gland condenser pressure to assure a vacuum is maintained. This will assure no water gets into the oil system.

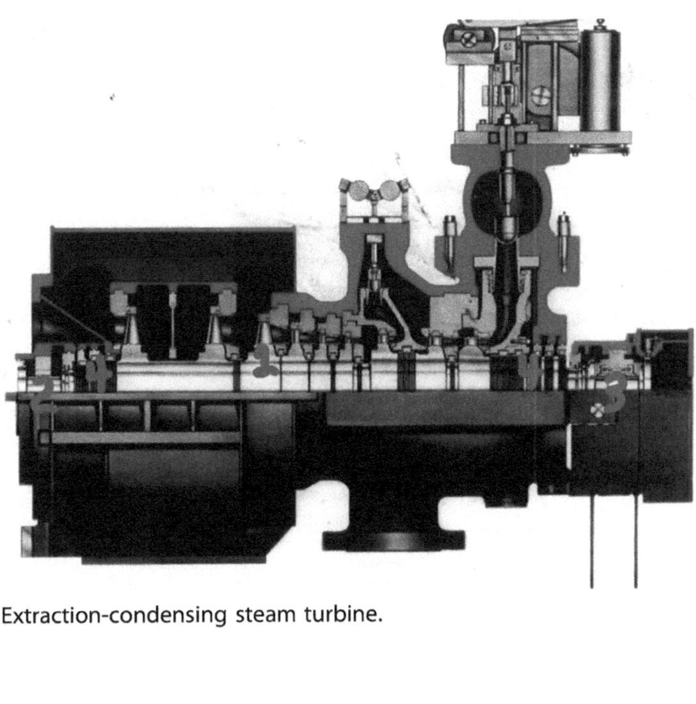

Fig. 1.8 Extraction-condensing steam turbine.

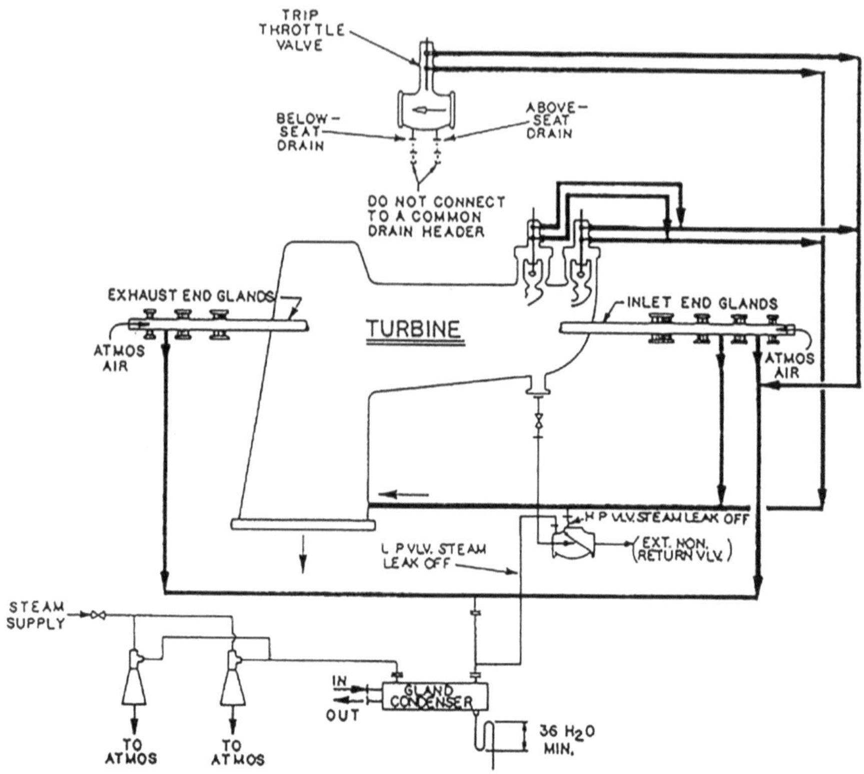

Fig. 1.9 Steam seal system schematic.

Gas Turbines

Without repeating myself again and making this book more boring than the subject already is, haha, I will just say that a gas turbine is a compressor and steam turbine in one machine. Therefore, the 1–4 components are the same as we have already discussed (see Fig. 1.10).

Due to the high temperatures seen in a gas turbine and the need for clean inlet air, however, there are a few more auxiliary systems. There is an air purge system to cool the hot turbine blades, a bearing air purge system, an air filtration system, a starting motor, and a start-up sequencing system. We will discuss these systems in more detail in Chapter 7, but at this point, all you need to know is there are quite a few more auxiliary system components that need to be monitored. The good thing is that gas turbines typically are highly instrumented and the OEM control panel contains all these parameters that can easily be trended.

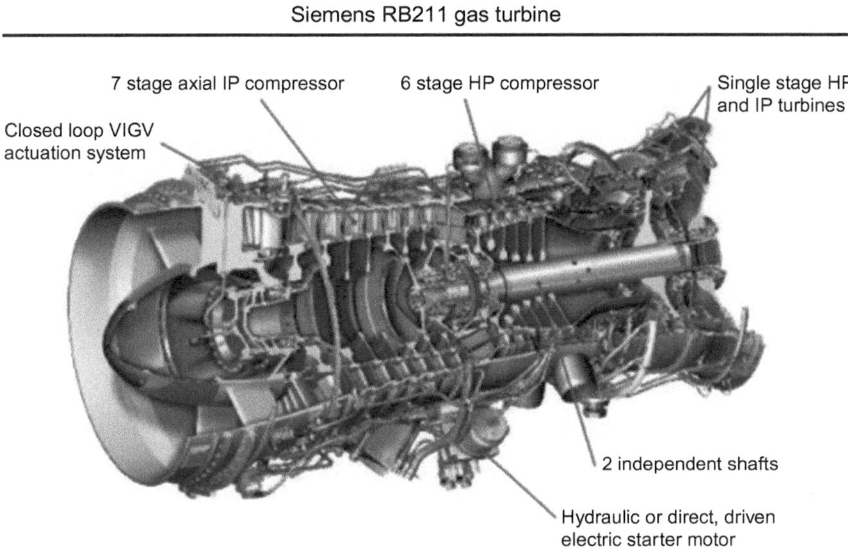

Fig. 1.10 Gas turbine with the major components.

Now we know the basics of what CCM is and that all types of rotating machinery can be divided into five major components. We will now get into the details of how to use CCM and what to monitor specifically for each of the components.

CHAPTER TWO

Before Not After—Utilizing a Baseline and Trends to Notice a Component Condition Change Prior to Alarm Setpoint

Chapter Outline

Now that we have covered the basics of what CCM is and the five major components for all rotating machinery, we will now discuss the practical use in everyday operation and the value added of this method of monitoring.

GET A BASELINE CONDITION

In order to know if a condition of any of the five major components has deteriorated, you need to know what the "good condition" was. This sounds like elementary, but the majority of the equipment (special purpose/unspared) are custom made for your specific application. Therefore, while there are guidelines of values that sound out of norm for all parameters, the normal field operating condition is slightly different for all machines. Field operating condition is stressed here, as the testing performed at the OEM facility prior to shipment never fully mimics the field conditions. Fig. 2.1 discusses the basics of when to capture the baseline condition.

As you probably figured out, the baseline is a snapshot of CCM at the beginning of a run for a machine (when all components are known to be in good condition). A Baseline condition that we use in everyday life is our regular medical physicals. A typical medical physical consists of having your height and weight documented, blood test, blood pressure, body temperature, hearing, and eyesight checked and an EKG to look at your hearts condition. The first time you have a physical exam, these parameters

Forsthoffer's Component Condition Monitoring
https://doi.org/10.1016/B978-0-12-809599-7.00002-6

When to capture your baseline conditions?

- At initial Start-up, and every start-up after rotor cleaning or change-out.
- After any component upgrade or rerate.
- If you have not captured a full CCM baseline, capture a full CCM during operation now and use this as a baseline until maintenance is performed next.
- Remember, the factory performance and mechanical run tests are not accurate field baselines, due to differences in field installation and amount and location of instrumentation. A field baseline differing from factory testing is absolutely normal!!!

Fig. 2.1 When to capture a baseline.

What to do when CCM differs (>10%) from baseline

- First, confirm all instrumentation is properly calibrated and/or not fouled.
- Confirm density of gas has not significantly changed (for performance monitoring in dynamic compressors)
  - If gas density changes more than 20%, new performance curves will need to be created by the OEM for the new gas composition.
- Use your CCM trending to determine the rate of degradation of component and the length of time you can operate until component alarm is reached.
- Plan for required maintenance during the most convenient time (hopefully an upcoming turnaround).
- Do not wait for alarms!!!

Fig. 2.2 Action to take when component condition change is observed.

are documented and filed away as your baseline condition. Your doctor then utilizes this information in future physical exams in order to see if any changes in conditions have occurred. CCM is just like a physical exam of rotating machinery and we need our baseline condition to compare future CCM snapshots (or our trended parameters) to.

This baseline should cover all components (rotor, journal bearings, thrust bearings, and auxiliary systems) at essentially the same time. If the unit (driver, driven, and transmission device) is instrumented properly, the majority of the component condition can be monitored in real time; therefore the baseline may essentially be able to be captured by observing your trends.

The key, though, is how we use the baseline along with our CCM trending in order to determine a component(s) issue. Fig. 2.2 describes the action to take when a condition change is observed.

The toughest thing to do when you see a machine, or component, degrading over time is to convince management (rightfully so) to plan for maintenance. Therefore, you need to be 100% sure that your observations are real. As described in Fig. 2.2, this consists of making sure all instrument is calibrated and/or replaced and there are no fouled pressure taps or thermowells. If that is confirmed, the next step is to determine the rate of degradation based on your trending that will allow you to know when the component condition may affect production ($). By doing this, you can plan for maintenance when production will begin to be affected, and management will always listen at that point.

BEFORE, NOT AFTER

Failure to act on a condition change immediately can result in an unplanned shutdown, plain and simple. Fig. 2.3 discusses the importance of this.

There is never enough manpower to constantly be reviewing trends, but if you set threshold limits to give the appropriate personnel warning of a component condition change, you can then have ample time to investigate the matter before it hits alarm.

A very common example of this is for a hydrodynamic journal bearing, where you may see initial field vibration as low as 10 μm (about 0.4 mils), and the alarm may be 50 μm (about 2 mils) or higher. If the vibration increases to 15 μm (about 0.6 mils), while it is nowhere near alarm, it has increased by 50%!!! That is significant and should be investigated as soon as possible. It could be something as simple as a process condition change that affected where a centrifugal compressor is operating on its curve, which will show up in monitoring the condition of the rotor. Can you now begin to see the importance of trending all five major components at the same time?

Fig. 2.4 discusses typical threshold limits for components.

It is very important that these limits are set up so that the machinery and/or reliability engineers are notified when the threshold is reached, not operations. Operations personnel have enough alarms to monitor every second

Before, not after

- By comparing your trends to the baseline, you can immediately see when the condition of one of the five major components has changed.
- A change of more than 10% should constitute investigation into the root cause.
- Set threshold limits on your trends to indicate when a significant change in condition is occurring.
- If you wait until alarm is reached, it may be too late!!!

Fig. 2.3 Before, not after.

Threshold limits by component

- Rotor—10 % change in performance from baseline
  - Head and flow for dynamic type compressors and pumps
  - Flow for positive displacement type compressors and pumps
  - Efficiency for turbines (Gas turbines should be investigated when a 5 % efficiency change is observed).
- Journal bearings—50% change in vibration from baseline condition
- Thrust bearings—25% change in axial displacement from baseline condition
- Seals—10 % change in leakage from baseline condition
- Auxiliary systems—10% change control valve positions from baseline condition

Fig. 2.4 Typical threshold limits by component.

of the day and adding these to the DCS will only be a nuisance to them. They should be notified of the findings of the investigation as to why the component condition change occurred.

Now, in Chapters 3–7 we will discuss the specific parameters that need to be monitored and trended in order to provide the most accurate CCM.

CHAPTER THREE

The Rotor

Chapter Outline

We will now focus on the five major components in the next five chapters of this book, with this chapter detailing the rotor for all types of rotating machinery. We will specifically discuss the parameters to monitor and typical limits for those parameters.

We define the rotor essentially as the component containing the pressure producing or power producing elements. In dynamic machinery and rotary positive displacement machinery, these are the rotating elements. For reciprocating machinery, however we define the rotor as the pressure producing elements (pistons, plungers, or diaphragms), which are moving in a lateral motion, connected to the rotating crankshaft.

We will now discuss the parameters to monitor for each specific type of machine.

POSITIVE DISPLACEMENT ROTORS

See Figs. 3.1 and 3.2 describing the principles of positive displacement (PD) vs. dynamic machinery.

As you can see in Figs. 3.1 and 3.2, positive displacement machines basically provide a constant volume flow for a given speed. Therefore, the most important parameter to be monitored and trended for all PD equipment is flow. If a flow decrease of at least 10% is observed (at a constant speed), this indicates internal components have worn to allow for internal bypassing of the process. At this point, maintenance should be planned for at a convenient time when the decreasing flow begins to affect plant rates.

Forsthoffer's Component Condition Monitoring
https://doi.org/10.1016/B978-0-12-809599-7.00003-8

Positive displacement Vs. dynamic

Positive displacement (screw, lobe, and reciprocating)

- Increases pressure by operating on a fixed volume in a confined space (constant volume if at constant speed).
- Need self-limiting device (relief valve or PSV) as flow will not change and pressure will therefore increase significantly.
- Not sensitive to changes within the system.

Dynamic (Centrifugal and Axial)

- Increases pressure by increasing fluid velocity with rotating blades and reducing the velocity in stationary components.
- Does not require self-limiting device as flow changes with pressure changes.
- Sensitive to changes on the system.

Fig. 3.1 Principles of positive displacement vs. dynamic machinery.

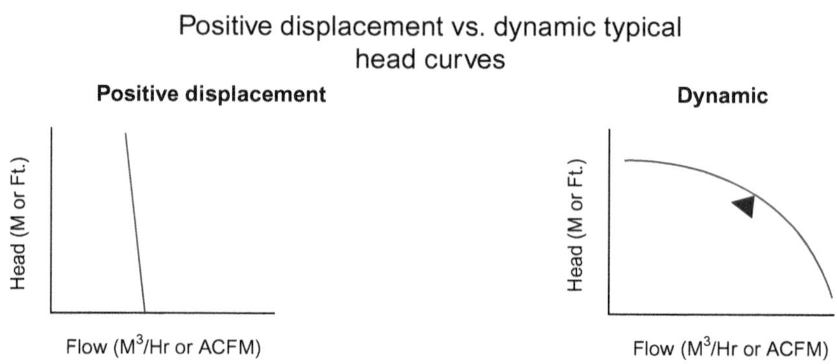

Fig. 3.2 Performance characteristics for positive displacement vs. dynamic machinery.

This is the case for all PD machines whether reciprocating piston, reciprocating plunger, reciprocating diaphragm, rotary screw, and rotary vane pumps or compressors. For reciprocating compressors however, another very important tool to monitor performance of the pressure producing elements (and components in contact with the pressure producing components) is by performing a data trap analysis. In order to do so, you need to have pressure taps (or strain gauges for pressures over 4500 psi) installed on the cylinders and a keyphasor on the crank shaft to determine the crank angle. This can be done portably or you can have a permanent system connected to continuously capture these data. Fig. 3.3 discusses the components that can be monitored by gathering this data.

The most important information produced from the data trap analysis is the creation of a Pressure-Volume (P—V) diagram, which is just like capturing an EKG at a physical exam. The P—V diagram should be created early on in the run of each cylinder and be regularly checked (quarterly

A portable or permanent "data trap" can monitor the following reciprocating compressor components

- Valves
- Bearings
- Packing
- Piston ring/rider band
- Rod load
- Crosshead pin bushing
- Piston rod/crosshead connection

Fig. 3.3 Reciprocating compressor data trap analysis benefits.

for portable checks) or even continuously with permanent systems in order to see degradation. Fig. 3.4 is a sample snapshot of a P—V for a cylinder. Note that this needs to be compared with previous P—V diagrams in order to analyze the condition of the components in the cylinder.

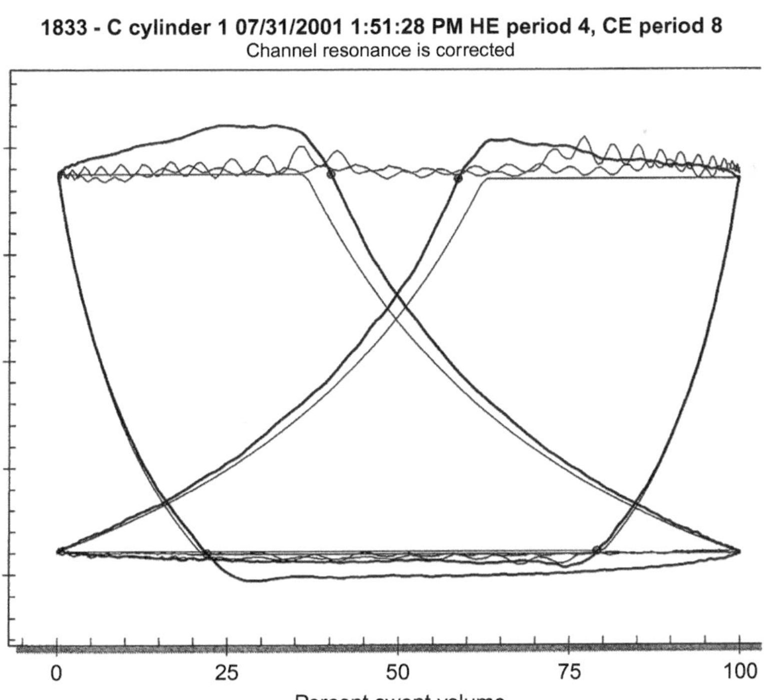

Fig. 3.4 Sample P—V diagram for a double acting cylinder.

Finally, adiabatic efficiency can be calculated and trended for all positive displacement machines. Table 3.1 is a spreadsheet calculation for adiabatic efficiency.

A decrease in efficiency by 10% would indicate wear and a plan for maintenance on the machine should be executed at the next convenience.

See Figs. 3.5–3.6 for a summary of the parameters to monitor and threshold limits for PD pump and compressor rotors.

Table 3.1 Adiabatic Efficiency Calculation

PD Compressor Performance Monitoring

Item #		
Date		
M.W.		
P_1 (kPaa)		
T (C)		
P_2 (kPaa)		
T2 (C)		
K		
Z		
Inlet flow (m^3/h)		
T2 Adiab. (C)	–	–
(K-1)/K	–	–
Gas density (kg/m^3)	–	–
Gas mass flow (kg/h)	–	–
Motor Amps		
Voltage		
Power factor		
Motor efficiency		
KW	–	–
Adiab. Hd (kj/kg)	–	–
Adiab. Hd (m)	–	–
Adiab. Eff'y	–	–
Does compressor need maintenance?		

Dry Screw Compressor Performance Monitoring

Item #	
Group	
M.W.	
P_1 (PSIA)	
T_1 (F)	
P_2 (PSIA)	
T2 (F)	
K	

Table 3.1 Adiabatic Efficiency Calculation—cont'd
PD Compressor Performance Monitoring

Z		
Inlet flow (ACFM)		
T2 adiab.	–	–
(K-1)/K	–	–
Gas density (lbs/ft^3)	–	–
Gas mass flow (lbs/min)	–	–
Motor Amps		
Voltage		
Power factor		
Motor efficiency		
BHP	–	–
Adiab. Hd (ft-lbf/lbm)	–	–
Adiab. Eff'y	–	–
Does compressor need maintenance?		

Rotary positive displacement rotor—
parameters to monitor

- Flow—A decrease in flow indicates internals wearing. A 10 % decrease in flow is a warning sign that a plan should be developed to perform maintenance at a convenient time.
- Adiabatic efficiency—Can be calculated (see Table 3.1) assuming you have proper instrumentation and can calculate/measure driver HP. A 10 % drop in efficiency is an indicator of internal components wearing and should be a warning sign that a plan should be developed to perform maintenance at a convenient time.
- It is always best to trend both parameters listed above to confirm that an instrument is not faulty!!!

Fig. 3.5 Parameters to monitor for rotary positive displacement rotor condition.

Reciprocating positive displacement rotor—
parameters to monitor

- Data Trap Analysis—Compare P-V charts to previous ones in order to see any change. Can be performed on a regular or Continuous basis by several companies.
- Flow—A decrease in flow indicates internals wearing. A 10 % decrease in flow is a warning sign that a plan should be developed to perform maintenance at a convenient time.
- Adiabatic efficiency—Can be calculated (see Table 3.1) assuming you have proper instrumentation and can calculate/measure driver HP. A 10 % drop in efficiency is an indicator of internal components wearing and should be a warning sign that a plan should be developed to perform maintenance at a convenient time.
- It is always best to trend as many parameters as possible for the same component to confirm that an instrument is not faulty!!!

Fig. 3.6 Parameters to monitor for reciprocating compressor rotor condition.

CENTRIFUGAL PUMP ROTORS

Fig. 3.7 shows a cross section of a multistage centrifugal pump with the rotor highlighted.

In order to check the condition of a centrifugal pump rotor, you will need to measure and/or calculate the following:

1. Suction pressure
2. Discharge pressure
3. Flow
4. Speed (if variable speed drive)
5. Head required $= \frac{.102 \times (P2 - P1)}{S.G.}$ in meters and pressure in bar, $\frac{2.311 \times (P2 - P1)}{S.G.}$ in feet and pressure in psi.

Note that for the head required, you will need to know the process fluid-specific gravity (S.G.) at the operating temperature. If you are going to assume the S.G. from the datasheets, you need to confirm that the process temperature is very close to the design process temperature.

Next, you need to compare the current operating point with the shop test performance curve (NOT PREDICTED!!!) to see the proximity of the operating point. Fig. 3.8 is a sketch of a centrifugal pump performance curve indicating this.

In order to plot your point, you need to calculate head and measure flow through the pump. If the measured head and flow fall at either point 1 or 2

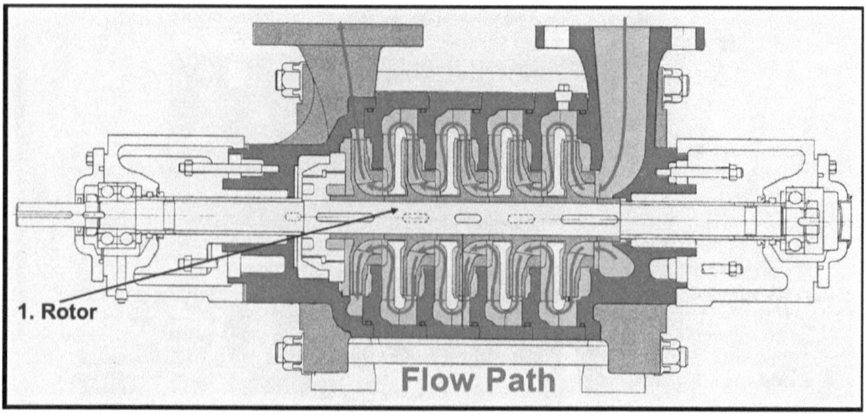

Fig. 3.7 Multistage centrifugal pump rotor.

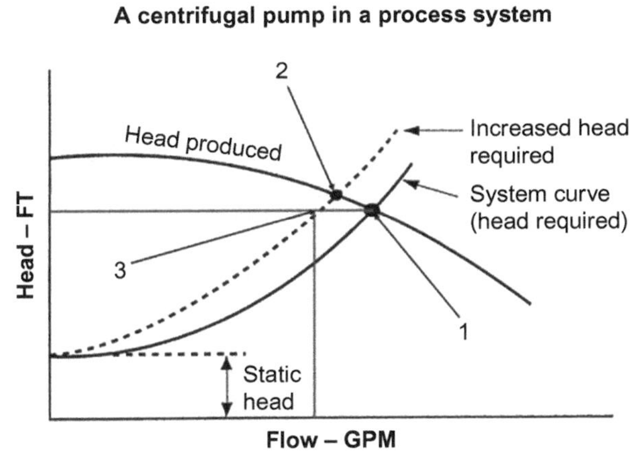

Fig. 3.8 Centrifugal pump performance curve sketch with different operating points.

on Fig. 3.8, this indicates the pump is in good condition. If, however, for a calculated head, the measured flow is less than that on the performance curve (like point 3 on Fig. 3.8), this could indicate internal component wear (Most Likely Wear Rings). Use caution though, if you have not performed a baseline check as many times the operating point does not fall exactly on the shop test performance curve. This is due to the fact that the instruments and process environment are not the same in the field as they are in the vendor shop. Therefore, the first field check of performance should always be your baseline to compare future checks. The threshold limit for centrifugal pump performance is a 10% decrease in flow for a given head required (see red line extending from point 3 to the performance curve in Fig. 3.8).

Another key centrifugal pump rotor condition parameter to check is called the equipment reliability operating envelope (EROE). This is an operating range (in terms of flow) that assures you are staying away from producing vapor in the pump (whether recirculation or cavitation) and you have minimal internal component wear. See Fig. 3.9 describing this on a performance curve.

Fig. 3.9 defines EROE as the operating range between −50% and + 10% of the flow at the best efficiency point for the pump, not necessarily the rated design flow. If you always operate within this range, all of the five major components in the pump will operate with optimal life. Depending on the type of pump, however this range may become narrower. For example, a centrifugal pump with an inducer has an EROE of approximately −10% to +10% of the BEP and pumps with low NPSHR (like double suction pumps)

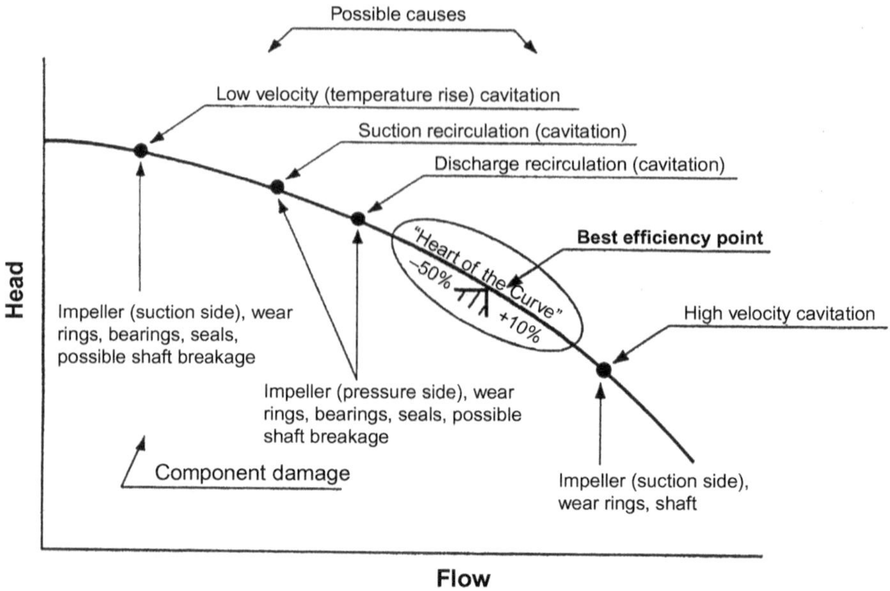

Fig 3.9 EROE

can have a smaller operating range than what is highlighted in Fig. 3.9. You will also see in Chapter 6 that EROE is the most important item to check to assure the mechanical seal is given its best chance for a long trouble-free life.

Therefore, the easiest way to determine what your EROE is for each of your pumps is to use the eye test and look at the performance curve (see Fig. 3.10).

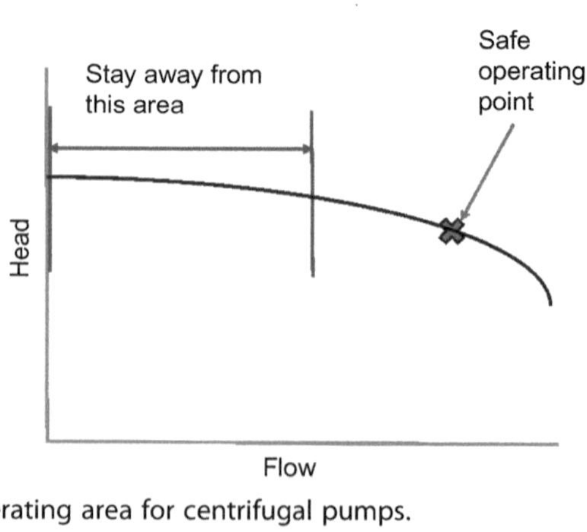

Fig. 3.10 Safe operating area for centrifugal pumps.

As you can see in Fig. 3.10, you want to keep your pump operating in the area of the curve where the slope is steep. In other words, a change in head required results in a small change in flow. When the curve gets flatter, a small change in head results in a large change in flow and is unstable. When a head increase of 3% or less results in zero flow (shutoff), that is the unstable region.

Not discussed so far has been the high range of EROE, which is where cavitation can occur. If your driver has enough power, and your process design has enough NPSHA, you can operate your pump beyond the +10% barrier safely. If either of those two factors are not valid, then the +10% side of the envelope needs to be observed.

Finally, for centrifugal pump rotors, we do not usually calculate and trend efficiency. The reason for this is that due to liquid being incompressible, there is typically very little temperature rise across the pump and a small inaccuracy of measuring temperature of the process liquid can result to a large change in efficiency calculated. Therefore, it is unable to be trended because the efficiency may change from one point to the next just due to how the user holds the infrared thermometer on the piping. Temperature rise across the pump is still useful though, but more in a practical manner. Typically, the temperature rise will be minimal (less than 5 °C) as long as the pump is operating within its EROE. However, if the pump is operating at a low efficiency (which drops off significantly at flows below the EROE), the liquid will recirculate more in the pump and heat up. So, Temperature rise can be used to check quickly if your pump is operating at a safe point on the curve. Also, if you are operating two pumps in parallel, the pump operating with more temperature rise across it is operating at a lower flow, meaning it is in worse condition than the other pump.

See Table 3.2 which is a spreadsheet that details all of the parameters needed to monitor and trend a centrifugal pump rotor (including calculations).

Note on the spreadsheet the other options to determine flow rate through the pump. As commonly found in the field, many pumps do not have individual flow meters, so you will have to be a little creative to determine what the flow is. The most accurate way to calculate flow is through a process control valve for the specific pump in question. You will be able to obtain the valve CV (flow coefficient) from the valve datasheets that the I&E group should have, and you will also need to know the valve position and pressure upstream and downstream of the valve (this should be available). If there is not an individual control valve for each pump, then you can calculate power from motor amps (if motor driver) and use the performance curve to

Table 3.2 Centrifugal Pump Performance Monitoring

Performance				
Item #				
Date				
Input				
$P_{2\ (psig)}$				
S.G.				
Pump speed (RPM)				
Flow rate (gpm)				
Calculate				
Head (ft)	–	–	–	–
Flow determination[a]				
Control valve position				
CV for valve and trim type[b]				
Fluid S.G.				
Control valve ΔP				
Calculated valve flow	–	–	–	–
Motor Amps				
Volts				
Power factor				
Motor eff'y				
Calculated power	–	–	–	–
Flow from pump curve[c]				
T1 (Deg. F)				
T2 (Deg. F)				
CP (specific eat)				
Calculated head	–	–	–	–
Calculated pump eff'y	–	–	–	–
Flow from pump curve[c]				
Pump maintenance required[d]				
EROE determination				
BEP flow[e]				
EROE min flow	–	–	–	–
EROE max flow	–	–	–	–
Is Pump in EROE	–	–	–	–
EROE targets for operations				
Flow				
Amps				
Pump ΔT (measured on inlet and discharge pipes)				

[a]If a flow meter is not available for the pump flow will be determined by one of the alternative methods noted in Table 7 of the supplementary manual. These alternative methods are: (1) Portable ultrasonic flow meter, (2) Control valve position to calculate flow, (3) Motor Amps to calculate power, (4) Pump Δ T to calculate pump efficiency.
[b]The CV can be found on the valve manufacturers curve for specific valve type and trim. Note that most of the major valve manufacturers post these curves on their websites.
[c]Flow will be estimated using the original shop test curve. Note that if the pump is not in good condition, these estimates will not be very accurate.
[d]Pump maintenance should be considered if the head and flow (operating point) when plotted on the test curve are approx. 10% below test curve flow for calculated head.
[e]Flow at highest efficiency taken from the shop test curve in the supplementary manual.

Centrifugal pump rotor—parameters to monitor

- EROE—determine if you are operating within the preferred operating range of flow. If not discuss with operations what they could do to increase the head required (increase P1, decrease P2, or increase S.G.) or decrease the head required (decrease P1, increase P2, or decrease S.G.) to get the flow back into the EROE.
- Plot operating point on performance curve—if the flow, for a given head required is 10% less or more from the curve, maintenance should be planned at next convenience.
- Use temperature rise to determine where the pump is in relation to EROE —if temperature rise from suction to discharge nozzles is greater than 5 degrees Celsius, the pump is operating at a lower efficiency and most likely below the minimum EROE flow.
- Always check instruments before developing an action plan!!!

Fig. 3.11 Parameters to monitor for centrifugal pump rotor.

determine the approximate flow rate. Finally, you can also use temperature rise (as mentioned earlier in this section) across the pump to give an idea of where the pump is running relative to its EROE.

See Fig. 3.11 which summarizes the parameters to monitor for a centrifugal pump rotor and the threshold limits that should indicate when action should be taken to determine what is causing the deviation.

The more these parameters and calculations can be brought into the DCS or PI and continuously monitored the easier it will be to determine a slight change in rotor condition. It is highly recommended to have all critical and bad actor pumps equipped with pressure and flow transmitters for this reason. The small investment will pay itself back as soon as you catch an issue early, saving the plant from prolonged troubleshooting and minimizing maintenance downtime.

CENTRIFUGAL COMPRESSOR ROTORS

Monitoring the rotor for a centrifugal compressor is very similar to a centrifugal pump. The only difference is the fact that gas is compressible and generates heat, therefore you can accurately calculate and trend efficiency. Also, because we are increasing the energy (head) of a vapor, we do not have to worry about the process fluid vaporizing, so there is not a specific flow range to operate in (other than staying away from surge and stonewall, wen applicable) like EROE.

See Fig. 3.12 for a picture of a large compressor rotor from a refrigeration service in an LNG plant.

Fig. 3.12 Centrifugal compressor rotor.

The key parameters to monitor and trend for a centrifugal compressor rotor are as follows:

1. Polytropic head – need P1, P2, T1, T2, gas analysis
2. Flow
3. Speed (if variable speed driver)
4. Efficiency
5. Horsepower required from compressor

These should be trended on the same time basis so you can truly see if an issue is occurring. For example, if flow goes down at a constant speed, you should see polytropic dead increase. If head does not increase in this example, this indicates internal wearing or possible fouling. Also, horsepower required should be increasing and efficiency should be decreasing.

See Table 3.3 which details all the parameters and calculations needed to monitor a centrifugal compressor rotor condition.

The calculations in Table 3.3 can be brought into your DCS or PI systems and plotted on the same time basis in real time. This will allow for you to catch a small change in condition before it turns into a loss of production.

Table 3.3 focuses on straight through compressors and/or compressors with intercoolers. In cracker units and LNG trains, as well as other processes, you will have refrigeration compressors that will contain multiple side-

Table 3.3 Centrifugal Compressor Rotor Performance Parameters and Calculations

Compressor Performance Monitoring

Item #				
Date				
M.W.				
P_1 (PSIA)				
T_1 (F)				
P_2 (PSIA)				
T_2 (F)				
K				
Z				
Inlet flow (ACFM)				
N (RPM)				
Calculate				
(K-1)/K	–	–	–	–
(n-1)/n	–	–	–	–
Gas density (lbs/ft^3)	–	–	–	–
Mass flow (lbs/min)	–	–	–	–
GHP	–	–	–	–
Poly Hd (ft-lbf/lbm)	–	–	–	–
Poly eff'y	–	–	–	–
Does compressor need maintenance?				

Compressor Performance Monitoring

Item #					
Date					
Given					
M.W.					
P_1 (bara)					
T_1 (C)					
P_2 (bara)					
T_2 (C)					
K avg					
Z1					
Flow (m^3/hr)					
N (RPM)					
Calculate					
(K-1)/K	–	–	–	–	–
(n-1)/n	–	–	–	–	–
Gas density (kg/m^3)	–	–	–	–	–
Mass flow (kg/hr)	–	–	–	–	–
KW	–	–	–	–	–
Poly Hd (kj/kg)	–	–	–	–	–
Poly Hd (m)	–	–	–	–	–
Poly eff'y	–	–	–	–	–
Does compressor need maintenance?					

Continued

Table 3.3 Centrifugal Compressor Rotor Performance Parameters and Calculations—cont'd

Compressor Performance Monitoring Exercise

Item #				
Group				
Given				
M.W.				
P1 (KPaa)				
T1 (C)				
P2 (KPaa)				
T2 (C)				
K				
Z				
Inlet Flow (m^3/hr)				
N (RPM)				
Calculate				
(K-1)/K	-	-	-	-
(n-1)/n	-	-	-	-
Gas density (kg/m^3)	-	-	-	-
Mass flow (kg/hr)	-	-	-	-
KW	-	-	-	-
Poly Hd (kj/kg)	-	-	-	-
Poly Hd (m)	-	-	-	-
Poly eff'y	-	-	-	-
Does compressor need maintenance?				

Compressor Performance Monitoring

Item #					
Group					
Given					
M.W.					
P_1 (kg/cm^2a)					
T_1 (C)					
P_2 (kg/cm^2a)					
T_2 (C)					
K avg					
Z1					
Flow (m^3/hr)					
N (RPM)					
Calculate					
(K-1)/K	-	-	-	-	-
(n-1)/n	-	-	-	-	-
Gas density (kg/m^3)	-	-	-	-	-
Mass flow (kg/hr)	-	-	-	-	-
KW	-	-	-	-	-
Poly Hd (kj/kg)	-	-	-	-	-
Poly Hd (m)	-	-	-	-	-
Poly eff'y	-	-	-	-	-
Does compressor need maintenance?					

stream inlets and/or outlets. These types of compressors are harder to monitor and trend performance, because there is no instrumentation to measure temperature at the discharge of each section (before a sidestream enters the case). The efficiency and head can still be estimated and trended using an iteration method. See Table 3.4, which is a spreadsheet containing calculations and instructions on how to perform this iteration.

Table 3.4 Refrigeration Compressor With Sidestreams Performance Calculations
Ethylene Refrigeration Compressor Performance Monitoring

	Section 1	SS-1	Section 2	SS-2	Section 3
Date					
Given					
M.W.					
P_1 (PSIA)					
T_1 (F)			–		–
P_2 (PSIA)					
T_2 (F)					
K avg					
Z1					
Inlet flow (ACFM)	–	–	–	–	–
Mass flow (lbs/hr)					
(K-1)/K	–	–	–	–	–
(n-1)/n	–	–	–	–	–
Gas density (lbs/ft^3)	–	–	–	–	–
Mass flow (lbs/min)	–	–	–	–	–
Gas horsepower	–	–	–	–	–
Poly Hd (ft-lbf/lbm)	–	–	–	–	–
Poly eff'y	–	–	–	–	–
Does compressor need maintenance?					

Color Codes for Values

	Fixed input values	Only check when output values indicate possible performance drop
	Values from DCS or PI	Dump values directly from DCS or PI
	Output values to trend	Trend these on same chart over time
	Estimated T2	Manually iterate this value to meet the required efficiency from performance curves
	Actual T2	This value should match what comes from Aspen, but for the first iteration plug in a value that gives you an efficiency close to what you read on the performance curve. Then continue to iterate all T2's until the actual T2 is met

Centrifugal compressor rotor—parameters to monitor

- Polytropic head vs. flow—trend and if flow drops by 10% or more for a constant head and speed, this indicates internal wear or fouling.
- Efficiency—A drop in efficiency by 10% can indicate internal wear or fouling of the compressor.
- Horsepower—an increase of HP required along with an efficiency drop will further indicate internal wear or fouling.
- When it appears there can be internal wear, first confirm all instruments are operating correctly and check that gas analysis has not changed!!!

Fig. 3.13 Parameters to monitor and threshold limits for a centrifugal compressor rotor.

The key at the bottom of Table 3.4 gives you directions on how to perform this calculation. As you iterate the discharge temperatures in each section to meet the actual overall discharge temperature, it is best to change the efficiency in each section by the same amount. While this is not an exact measurement of efficiency for each section, you will notice an overall drop in efficiency for the machine when it occurs. Also, remember if you perform the same calculation over time and trend the results, you will be able to see a condition change which is ultimately the purpose.

See Fig. 3.13 outlining the key parameters to monitor for a centrifugal compressor rotor and threshold limits.

STEAM TURBINE ROTORS

Steam turbine rotors are just like a dynamic (centrifugal and axial) compressor rotor with the exception that they take energy out of the gas (steam) rather than adding energy to it. See Fig. 3.14 showing a multistage steam turbine rotor.

Also, there is no performance test done on steam turbines, since the power produced is determined by the steam properties and flow rate. Therefore, field operating conditions are compared to the theoretical steam map generated by the OEM.

The key parameters to monitor and trend for a steam turbine rotor are as follows:

1. Inlet steam temperature
2. Inlet pressure
3. Exhaust temperature
4. Exhaust pressure

Fig. 3.14 Special purpose steam turbine rotor.

5. Inlet steam flow
6. Efficiency
7. After 1st stage pressure
8. Throttle valve opening %

In order to calculate efficiency, you need to use a steam table or Mollier diagram (see Fig. 3.15) in order to determine the enthalpy, or energy in the steam at the inlet and outlet conditions.

You then calculate the theoretical steam rate (the amount of steam flow to produce 1 unit of power at 100% efficiency) and actual steam rate (the amount of steam flow that is actually being required to produce 1 unit of power). The efficiency is then calculated by dividing the theoretical steam rate (TSR) into the actual steam rate (ASR). It sounds simple, yet gets complicated especially for condensing steam turbines, since you need to know the moisture in the exhaust steam in order to determine the energy at the exhaust, and there is not a real accurate way to measure the % moisture. So, for a condensing steam turbine, the easiest way to determine the actual steam rate is to know the power being provided by the turbine. This can be measured with a torquemeter, if equipped, or by using the power calculated that is being required by the compressor (discussed earlier in this chapter, and equation is embedded in Table 3.3) and add in an estimated power loss for bearings and seals. See Fig. 3.16 for more details on calculating steam turbine efficiency.

The calculated efficiency needs to be trended and compared to its baseline calculation. If it drops by more than 10%, this indicates internal wearing

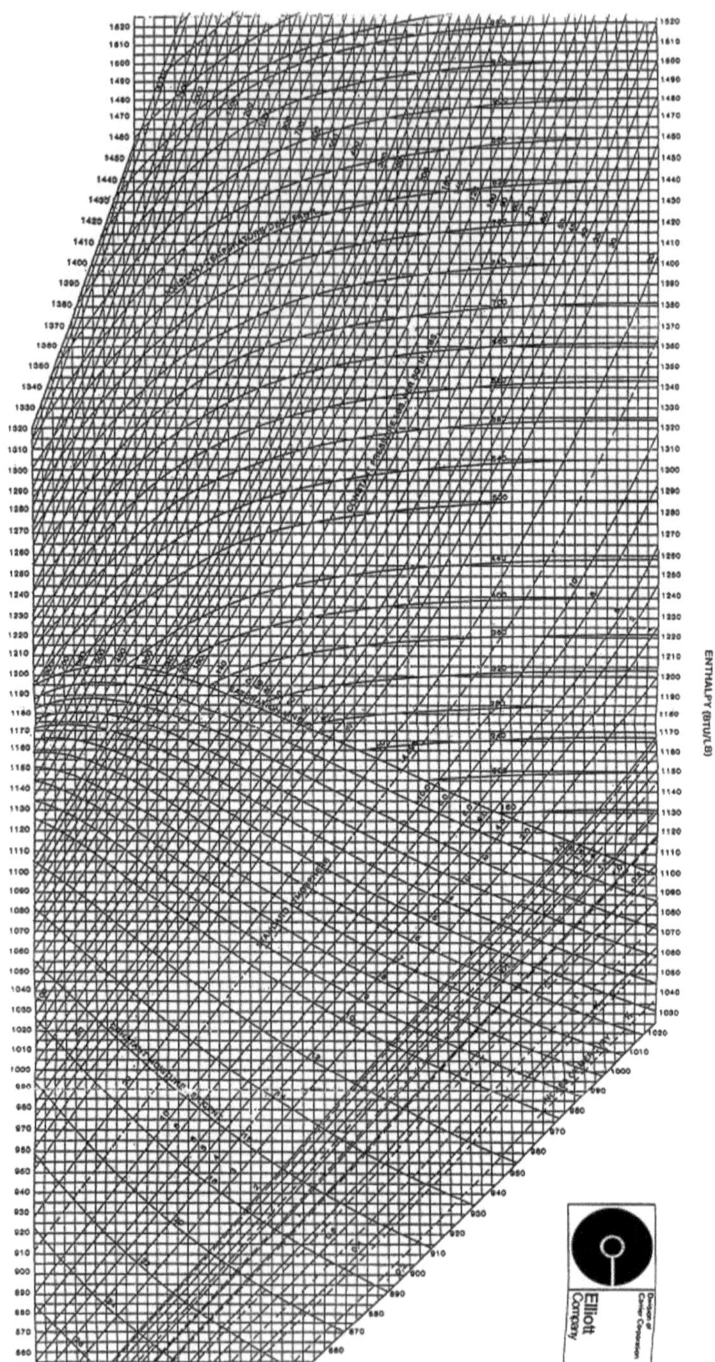

Fig. 3.15 Mollier diagram. *Elliott Company*

Determining steam rate

Uses:
- Determine the amount of steam required per hour
- Determine the amount of potential KW (horsepower)

Required:
- Steam conditions
- Theoretical steam rate table or Mollier diagram
- Thermal efficiency of turbine

Formula:

Metric units

- Theoretical steam rate

$$\text{T.S.R. (kg/kW-hr)} = \frac{3600 \text{ kJ/kW-HR}}{\Delta H_{\text{ISENTROPIC}}}$$

- Actual steam rate

$$\text{A.S.R. (kg/kW-hr)} = \frac{\text{T.S.R.}}{\text{Efficiency}} = \frac{3600 \text{ kJ/kW-HR}}{\Delta H_{\text{ACTUAL}}}$$

- Turbine efficiency

$$\text{Efficiency} = \frac{\text{T.S.R.}}{\text{A.S.R}} = \frac{\Delta H_{\text{ACTUAL}}}{\Delta H_{\text{ISENTROPIC}}}$$

U.S. units

$$\text{T.S.R. (lb/HP-hr)} = \frac{2545 \text{ BTU'S/HP-HR}}{\Delta H_{\text{ISENTROPIC}}}$$

$$\text{A.S.R. (lb/HP-hr)} = \frac{\text{T.S.R}}{\text{Efficiency}} = \frac{2545 \text{ BTU/HP-HR}}{\Delta H_{\text{ACTUAL}} \text{HP/HR}}$$

= Inlet steam flow/ actual HP

Fig. 3.16 Calculating steam rate for steam turbine performance.

or fouling of the blades. Note, however that you want to be monitoring and trending the throttle valve opening as well. If there is plenty of travel left in the throttle valves, you do not have to plan for maintenance immediately but determine the rate of deterioration and when you will need to plan for maintenance.

Tables 3.5 and 3.6 are spreadsheets providing the parameters and calculations needed to calculate steam turbine efficiency for a backpressure and extraction-condensing turbine respectively. For a straight condensing turbine, you can just use the LP section of Table 3.6.

Probably the most important parameter to monitor for steam turbine rotor condition is the after first stage (and after extraction stage on extraction-condensing turbines) pressure. This tells you the amount of steam pressure drop across the first row of blades, and all OEMs produce a curve like that shown in Fig. 3.17.

On the *x*-axis is inlet steam flow and after first stage pressure on the *y*-axis. For a given flow rate the first stage pressure should be on the curve. If it is above the curve, this indicates that not enough pressure (energy) has been

Table 3.5 Back Pressure Steam Turbine Efficiency Calculations

Backpressure Steam Turbine Performance

Item #				
Date				
Given				
P_1 (kPaa)				
T1 (Deg. C)				
P2 (kPaa)				
T2 (Deg. C)				
Speed (RPM)				
Flow Rate (kg/hr)				
Determine				
[a]h1 (KJ/kg, from Mollier diagram)				
[a]h2 isentropic (KJ/kg, from Mollier diagram)				
Theoretical steam rate (kg/KWhr, TSR)	–	–	–	–
[a]h2 actual (KJ/kg, from Mollier diagram)				
Steam turbine efficiency	–	–	–	–
Actions required				

BackPressure Steam Turbine Performance

Item #				
Date				
Given				
P_1 (psia)				
T1 (Deg. F)				
P2 (psia)				
T2 (Deg. F)				
Determine				
[a]h1 (BTU/lb, from Mollier diagram)				
[a]h2 isentropic (BTU/lb, from Mollier diagram)				
Theoretical steam rate (lb/HPhr, TSR)	–	–	–	–
[a]h2 actual (BTU/lb, from Mollier diagram)				
Steam turbine efficiency	–	–	–	–
Actions required				

[a]Note that values for h1, h2 isentropic, and h2 actual (HP case only) are obtained from a Mollier diagram contained in this section or by using steam tables.

taken out of the steam, which can only be the case if the rotor blade passages have significant deposits on them, restricting steam flow. Now, if the pressure is below the curve for a given flow rate, this can indicate that the blades on the rotor have corroded, most likely from moisture in the back stages. Typically, this parameter is measured by a pressure gauge in the field, but

Table 3.6 Extraction-Condensing Steam Turbine Efficiency Calculations
Extraction-Condensing Steam Turbine Performance

	1		2	
Item #				
Group				
Given				
P_1 (kpa)				
T1 (Deg. C)				
P2 (kpa)				
T2 (Deg. C)				
Speed (RPM)				
Flow Rate (kg/hr)				
[a]Total Power at Coupling (KW)				
Determine				
[b]h1 (KJ/kg, from Mollier diagram)				
[b]h2 isentropic (KJ/kg, from Mollier diagram)				
Theoretical steam rate (kg/KWhr, TSR)	–	–	–	–
KW HP section	–		–	
KW LP section		–		–
[b]h2 actual (KJ/kg, from Mollier diagram)		–		–
Actual steam rate (kg/KWhr, ASR)	–	–	–	–
Steam turbine efficiency	–	–	–	–
% Moisture LP section				
Actions required				

[a]Total power at coupling can be determined via torque meter or from compressor total gas power plus mechanical losses for bearings and seals.
[b]Note that values for h1, h2 isentropic, and h2 actual (HP Case only) are obtained from a Mollier diagram contained in this section or by using steam tables.

it is highly recommended to install a transmitter to measure this pressure and have the ability to trend it. Remember, this parameter will tell you if the turbine is fouled or if it has lost its efficiency due to internal components wearing, allowing you to better plan for maintenance when it is needed.

Fig. 3.18 summarizes the key parameters to monitor a steam turbine rotor and the threshold limits.

Note that the discussion on steam turbine rotor monitoring has mainly been focused on special purpose steam turbines that drive large equipment. We typically do not calculate and trend efficiency on general purpose (single valve) steam turbines because they usually do not have inlet flow and all the steam pressures and temperatures. However, if you have all of the proper instruments, you can calculate the efficiency of these turbines using Table 3.4 (backpressure turbine efficiency calculations). The main item to

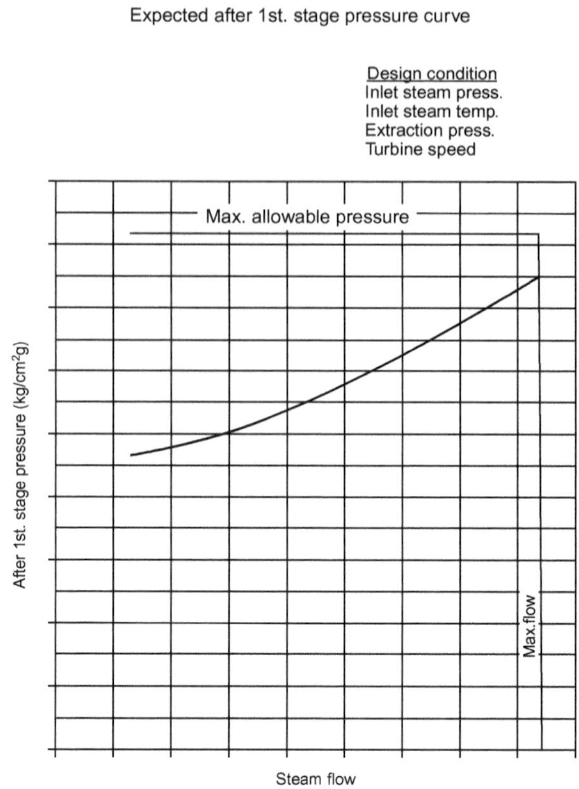

Fig. 3.17 After first stage pressure curve.

Steam turbine rotor—parameters to monitor

- Efficiency—trend and if it drops by more than 10%, plan for maintenance when throttle valve opening is greater than 90% open.
- After first stage/after extraction pressure—trend and if efficiency drops, plot current point on appropriate curve to see if efficiency loss is due to fouling or corrosion.

Fig. 3.18 Steam turbine rotor – parameters to monitor.

monitor and trend for general purpose turbine rotors is the governor opening to assure you have enough power available to drive your equipment over changing process conditions.

GAS TURBINE ROTORS

To monitor the performance of a gas turbine rotor is basically like combining what we have learned about monitoring compressor and steam turbine rotors. See Fig. 3.19 that shows an Aero-derivative type gas turbine.

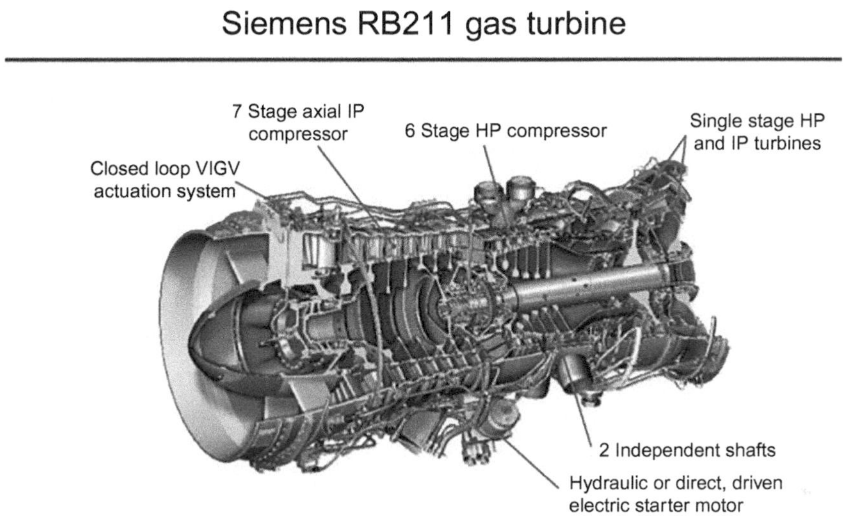

Fig. 3.19 Aero-derivative gas turbine.

As you can see, there is an axial compressor (sometimes split up into two sections like the RB211 depicted) that increases the air pressure in order to produce combustion when combined with the fuel in the combustion chambers. The energy generated from combustion then is transferred to produce power by the turbine on the back end.

Fig. 3.20 shows a typical gas generator rotor for a two-shaft industrial gas turbine.

Fig. 3.20 Gas generator rotor for two shaft industrial GT.

As can be seen in Fig. 3.20, the left portion of the rotor is the axial compressor portion and the one row of blades on the right are the gas turbine portion of this rotor that is used to generate enough power to drive the axial compressor. The hot gas leaving the gas turbine blades will then flow to a second rotor that contains the power turbine blades, used to drive your generator or large industrial rotating machine (usually compressors, but sometimes used with large multistage pumps).

The first set of parameters to monitor is for the axial air compressor portion, which is essentially the same as for a centrifugal compressor. Following are the parameters to monitor for the axial air compressor portion:

1. Head—need ambient pressure, P2, ambient temperature, T2, humidity
2. Flow
3. Speed (if variable speed driver)
4. Efficiency

The air compressor is the most critical portion of the gas turbine rotor to monitor as it has the greatest effect on overall machine efficiency. If the delivered pressure or volume of air to the combustion chambers is reduced, the energy created during combustion is reduced, producing less power. The degradation of the air compressor can be from damage due to the blow off valve not opening when required, but the most common form of degradation is from the forming of dirt, sand, and salt particles on the axial compressor blades. This will happen to every axial air compressor over time with the rate of this fouling differing depending on the ambient conditions where it is installed and the inlet filtration system effectiveness.

The most accurate way to monitor the condition of the axial compressor portion is by calculating and trending efficiency. However, you need compressor discharge temperature (CDT) to do this and many gas turbines are not equipped with this at installation. If you do not have CDT available, contact the vendor of the equipment as he or she usually has provisions to install a CDT transmitter easily. If you are just monitoring pressure ratio, you are not taking the gas density into effect and may see an effect on pressure ratio that is just caused by ambient condition changes.

The GT efficiency is also important to monitor (even though it is usually directly impacted by axial compressor performance) and trend. The GT efficiency is very similar to calculating steam turbine efficiency, the only difference is the fuel can differ and the combustion is produced internally not externally in a boiler like a steam turbine. Figs. 3.21 and 3.22 explain the differences between steam turbine and gas turbine performance including the calculations.

Gas turbine vs steam performance

- Since the steam turbine (Rankine) cycle is closed, the vapor and vapor conditions are constant!
- Therefore performance can be expressed in terms of:
- Steam rate = $\dfrac{\text{LBS of steam}}{\text{BHP or KW–HR}}$
- External efficiency = $\dfrac{\text{Actual work x loss factor*}}{\text{Ideal work}}$

* Includes mechanical and leakage losses

Fig. 3.21 Comparison between gas turbine and steam turbine performance.

- Since the gas turbine (Brayton) cycle is open, both the vapor and vapor conditions are variable.
- Therefore performance is expressed in terms of:
- Heat rate (ISO)*1 = $\dfrac{\text{BTU}}{\text{BHP or KW-HR}}$
- Thermal efficiency = $\dfrac{\text{BTU/HP-HR}}{\text{Heat rate (ISO) BTU/HP-HR}}$
- Fuel rate = $\dfrac{\text{(Heat rate) X (horsepower)}}{\text{Fuel heating value (BTU/LB)}}$ *2

Note: *1 ISO conditions—standardized fuel, inlet conditions at design speed—no losses

*2 BTU/SCF (standard cubic foot) for gas fuel

Fig. 3.22 Gas turbine vs. steam turbine performance continued.

Table 3.7 is a spreadsheet containing the rotor parameters to monitor and trend with calculations embedded. Again, these can be input into the DCS and usually most gas turbine control systems do include screens that contain the majority of these calculations and trending capabilities.

Note that for the GT efficiency, you will need to know the fuel flow and fuel lower heating value (LHV can be gathered from vendor datasheets or

Table 3.7 Gas Turbine Rotor Performance Calculations

	Gas Turbine Component Condition Monitoring
Equip. #	
Date	
Compressor performance	
P atmosphere (bara)	
T ambient (C)	
P Disch. (bara)	
T Disch. (C)	
Speed (RPM)	
K	
(K-1)/K	– – – – – – –
(n-1)/n	– – – – – – –
Polytropic efficiency	– – – – – – –
What action is required based on % difference from design efficiency?	
Gas turbine performance	
Fuel LHV (KJ/Kg)	
Flow (Kg/s)	
Output KW	
Site heat rate (KJ/KW-hr)	– – – – – – –
GT efficiency	– – – – – – –
What action is required based on gas turbine performance info. above?	

found online) as well as the power being required from the driven equipment (can be measured from a torquemeter or calculated from compressor performance).

Fig. 3.23 summarizes the parameters to monitor for a gas turbine rotor and the threshold limits.

Gas turbine rotor—parameters to monitor

- Axial air compressor efficiency—A drop in efficiency by 5% indicates fouling is most likely occurring. Perform online or crank wash as soon as possible!!!
- Gas turbine efficiency—5% or more drop in efficiency indicates components wearing, axial compressor fouling, or combustor issues. Try washing the compressor first, then plan for maintenance with vendor consulting if washing does not give an efficiency gain.

Fig. 3.23 Parameters to monitor for gas turbine rotors.

Note the comments in Fig. 3.23 on washing of the axial compressor. There are several different washing system setups and you can wash the machine online and at crank speed, each of which offer advantages. A very detailed article was presented at the 2013 Texas A&M Middle East Turbomachinery Symposium held in Qatar (METS), that discusses the benefits of washing and the different setups. This article is included in the appendix of the book.

Now, we will discuss radial bearing condition monitoring and continue with the other major components after that. Remember that just looking at 1 or 4 of the 5 major components can result in missing an issue and prolonged troubleshooting. Always monitor and trend all five components on the same timeline!!!

CHAPTER FOUR

Journal (Radial) Bearings

Chapter Outline

In this chapter we will focus on monitoring component # 2 of the five major components in rotating machinery, the journal (or radial) bearings.

There are only two types of bearings used among all the equipment types, antifriction and hydrodynamic.

ANTIFRICTION BEARINGS

An antifriction bearing is fitted to the shaft, usually with an interference fit and utilizes rolling elements, bound by inner and outer rings that absorb the radial forces exerted on the rotor. Antifriction-type bearings are typically applied to pumps and also aero-derivative gas turbines. Fig. 4.1 shows a centrifugal pump fitted with antifriction ball bearings.

Fig. 4.2 is a diagram of antifriction-type bearings.

Since an antifriction bearing is fitted to the shaft and has a tight clearance to the casing, when the shaft vibrates, the force is absorbed by the rolling elements and transmitted to the casing. Therefore, you monitor bearing housing vibrations and temperatures for an antifriction-type bearing. Table 4.1 is a spreadsheet showing the parameters to monitor for a pump equipped with antifriction bearings.

Typically, the parameters listed in Table 4.1 are taken on a monthly basis using a portable data logger (CSI, etc.), but it is highly recommended to have critical and bad actor pump casings equipped with permanent accelerometers, so the values can be continuously monitored and trended.

When vibration has increased by more than 20%, the vibration group should capture a spectrum analysis to analyze at what frequency the majority of the vibration increase occurred. This should give you an idea of what

Forsthoffer's Component Condition Monitoring
https://doi.org/10.1016/B978-0-12-809599-7.00004-X

Fig. 4.1 Pump with antifriction ball bearings.

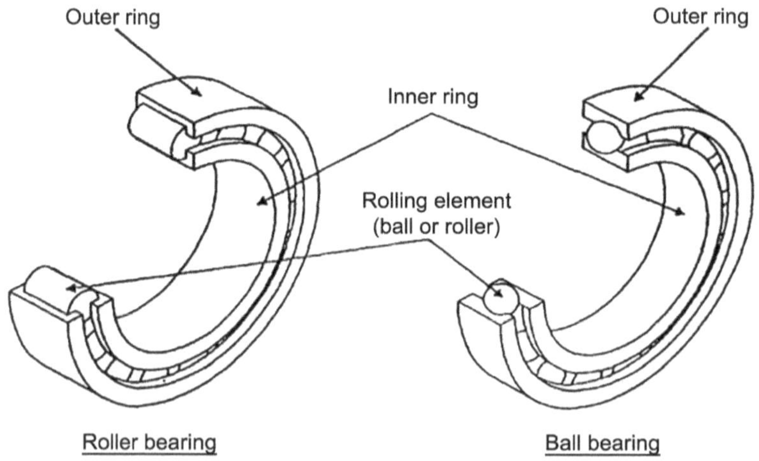

Fig. 4.2 Antifriction bearings.

changed to cause the vibration increase. See Figs. 4.3 and 4.4 which use the analogy of an AM/FM radio to discuss this diagnostic tool.

For pumps, or equipment with antifriction bearings, the main frequencies to look at are the 1 × (indicates piping or foundation forces), 2 × (indicates misalignment), or the number of rolling elements × the operating speed (indicates bearing failure).

Fig. 4.5 summarizes the parameters to monitor antifriction bearings and their threshold limits.

Table 4.1 Antifriction Bearing Monitoring

	Pump Bearing System Condition Monitoring
Item #	
Date	
Housing temperature (Deg F)	
Outboard journal	
Coupling end journal	
Housing vibration (in/s)	
Outboard journal horizontal	
Outboard journal vertical	
Coupling journal horizontal	
Coupling journal vertical	
Axial vibration	
Oil condition	
Appearance	
Water content	
Constant level oiler condition	
Bottom bracket oil condition monitor? (Y/N)	
Action required	

ANY VIBRATION SIGNAL IS MADE UP OF ONE OR MORE FREQUENCIES. A TYPICAL UNFILTERED (OVERALL) AND FILTERED SIGNAL ARE SHOWN BELOW:

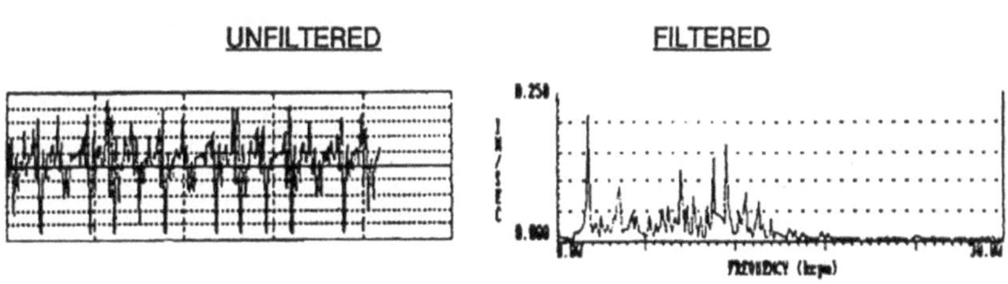

AN ANALOGY TO A FILTERED SIGNAL IS A RADIO. IN A GIVEN LOCALITY, MANY STATIONS ARE TRANSMITTING SIMULTANEOUSLY. ANY GIVEN STATION IS OBTAINED BY ADJUSTING THE TUNER TO THE CORRECT TRANSMISSION FREQUENCY.

Fig. 4.3 Vibration frequency.

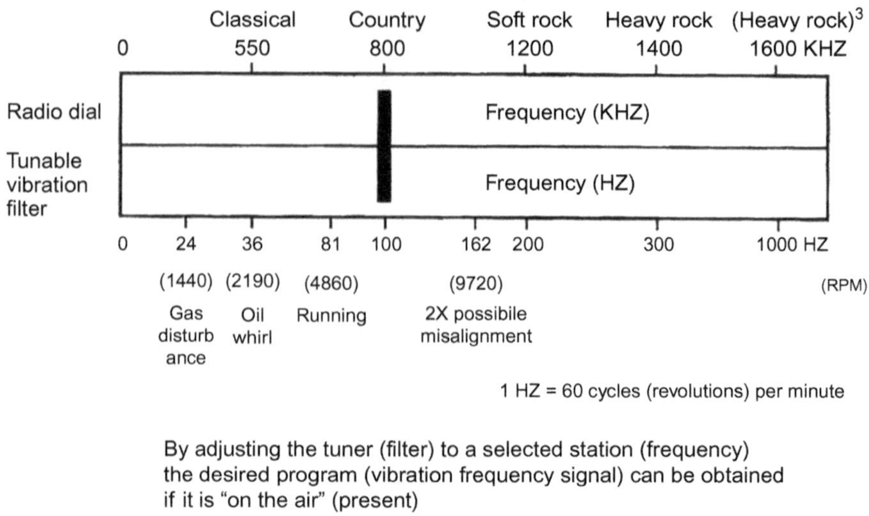

Fig. 4.4 Radio tuner analogy to vibration frequency.

Antifriction radial bearings—parameters to monitor

- Bearing bracket vibration—An increase by 20% of baseline should warrant a vibration frequency check to analyze what is causing the vibration increase.
- Bearing bracket temperature—If temperature increases by 20%, look at oil condition and water content in bearing bracket
- Remember to always look at performance on same timeline as vibrations as vibrations will change depending on where you are operating on the performance curve!!!

Fig. 4.5 Parameters to monitor for antifriction bearings.

HYDRODYNAMIC BEARINGS

Hydrodynamic bearings use oil to support the rotor. The bearing itself has a clearance (typically 1.5 thousandths of an inch per inch of shaft diameter) on the shaft, and oil is injected between the bearing and shaft. The oil wedge created (the oil between the bearing and shaft at the load point) is only about 20–25 μm tick, about the thickness of a human hair. Oil can handle a load of approximately 500 psi before it breaks down, so the equipment

Fig. 4.6 Hydrodynamic tilting pad journal bearing.

OEM selects a bearing design and size to be loaded at a maximum of 250 psi at normal operating conditions. See Fig. 4.6 for a picture of typical hydrodynamic tilting pad journal bearings which is used in the majority of compressors and steam turbines.

Since there is a clearance between the bearing and shaft for a hydrodynamic bearing, we can monitor the amount of vibration on the shaft. Therefore, equipment with these types of bearings are typically fitted with vibration probes (usually by Bentley Nevada/GE) that are charged with an AC current and use voltage gap to calculate how close the shaft is to the probe on every revolution. See Fig. 4.7 for a description of this setup.

The parameters that are typically monitored in real time for these bearings are overall vibration and bearing pad temperature. See Table 4.2 for a spreadsheet showing the parameters monitored and trended for hydrodynamic journal bearings in a steam turbine-driven compressor train.

When vibration or temperature increases by 20%, you then need to utilize the diagnostic systems in your plant to analyze the cause of the vibration increase. These days, the majority of compressor trains are equipped with Bentley Nevada System 1 or equivalent systems that can analyze phase angle, bearing orbit, and shaft centerline position. Fig. 4.8 details all of these parameters and typical alarm limits.

Again, remember that the parameters discussed in this chapter need to be trended with the other component condition parameters. For example, an

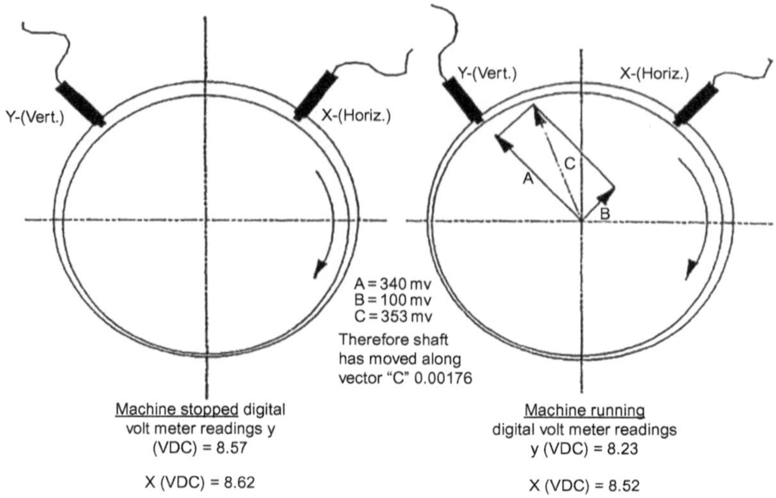

Fig. 4.7 Vibration probe setup.

Table 4.2 Hydrodynamic Journal Bearings Parameters to Monitor

	Steam Turbine/Compressor Train Journal Bearing Monitoring
Item #	
Date/Time:	
Journ. Brgs.	
Compressor DE horiz. vibes (micron)	
Compressor DE vert. vibes (micron)	
Compressor DE pad temp (Deg C)	
Compressor DE pad temp (Deg C)	
Compressor NDE horiz. vibes(micron)	
Compressor NDE vert. vibes (micron)	
Compressor NDE pad temp (Deg C)	
Compressor NDE pad temp (Deg C)	
Steam turbine DE horiz. vibes (micron)	
Steam turbine DE vert. vibes (micron)	
Steam turbine DE pad temp. (Deg. C)	
Steam turbine DE pad temp. (Deg. C)	
Steam turbine NDE horiz. Vibes (micron)	
Steam turbine NDE vert. vibes (micron)	
Steam turbine NDE pad temp. (Deg. C)	
Steam turbine NDE pad temp. (Deg. C)	

Hydrodynamic journal bearings—parameters to monitor

• Overall vibration	• Increase of 20% from baseline indicates to use diagnostic tools (typical alarm of 2.5 mils or 60 microns)
• Pad temperature	• Increase of 20% from baseline indicates to use diagnostic tools (typical alarm of 220F or 108C)
• Phase angle	• Increase of 30 degrees indicates shift of heavy spot—possible fouling—confirm with frequency of 1× increase.
• Shaft centerline position	• Increase of 30 % in direction of load indicates bearing wearing.
• Vibration frequency	• Use this tool to see at what frequency the vibration has increased. 1× = unbalance, 2× = misalignment, etc.

Fig. 4.8 Hydrodynamic journal bearings parameters to monitor.

increase in 1 × vibration, for a compressor that is in a service that could foul, indicates fouling, but needs to be confirmed by trending efficiency over the same time that the 1 × vibration has increased.

Also, components 2–4 (journal bearings, thrust bearings, and seals) depend on their support systems in order to operate properly for an acceptable lifetime; therefore it is essential to monitor and trend the support system parameters along with these components. Chapter 7 will discuss these systems in detail.

Now we will discuss the monitoring of antifriction and hydrodynamic thrust bearings which is not much different than monitoring journal bearings.

CHAPTER FIVE

Thrust Bearings

Chapter Outline

This chapter will focus on monitoring and trending parameters for Thrust Bearings.

Just like radial bearings, there are only two types of thrust bearings used among all the equipment types, antifriction and hydrodynamic.

ANTIFRICTION THRUST BEARINGS

Most single-stage pumps do not exhibit a lot of thrust, and are typically fitted with an antifriction bearing that can accept some thrust load. Fig. 5.1 shows a centrifugal pump fitted with antifriction ball bearings. Note the bearing closes to the coupling is wider than the one closest to the process. The wider bearing is what is called a double row angular contact ball bearing.

Fig. 5.2 shows a cutaway of this type of bearing to see how it works.

You can see in Fig. 5.2 that there are two rows of rolling elements (balls) and there is shoulder at the outer diameter for each of the rows of balls toward the center of the assembly. These shoulders allow the balls to absorb the thrust force in either direction, which is enough axial support for a single stage pump. Remember, this bearing is also supporting the rotor in the radial direction at the same time.

Table 5.1 should look familiar, as it is the same table used in Chapter 4 to outline the parameters taken on a monthly basis by the condition monitoring group. Axial vibration is the key parameter used to monitor the thrust component absorbed by the angular contact bearings.

Typically, the parameters listed in Table 5.1 are taken on a monthly basis using a portable data logger (CSI, etc.), but it is highly recommended to have

https://doi.org/10.1016/B978-0-12-809599-7.00005-1

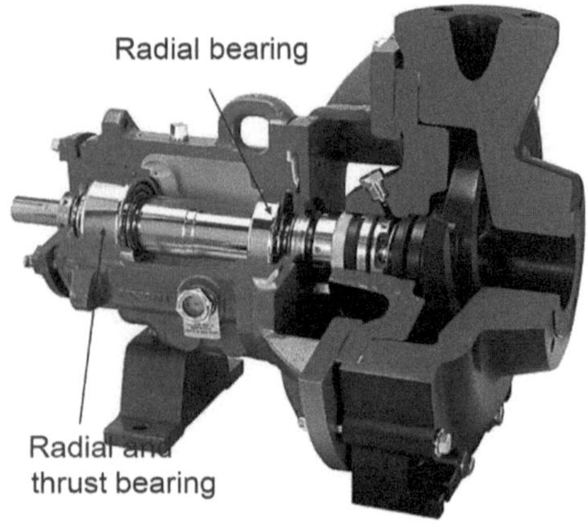

Fig. 5.1 Pump with antifriction ball bearings.

Fig. 5.2 Double row angular contact bearing. *(Courtesy of www.skf.com).*

Table 5.1 Antifriction Bearing Monitoring

	Pump Bearing System Condition Monitoring
Item #	
Date	
Housing temperature (Deg F)	
Outboard end	
Coupling end	
Housing vibration (in/s)	
Outboard journal horizontal	
Outboard journal vertical	
Coupling journal horizontal	
Coupling journal vertical	
Coupling end axial vibration	
Oil condition	
Appearance	
Water content	
Constant level oiler condition	
Bottom bracket oil condition monitor? (Y/N)	
Action required	

critical and bad actor pump casings equipped with permanent accelerometers, so the values can be continuously monitored and trended.

When vibration has increased by more than 20%, the vibration group should capture a spectrum analysis to analyze at what frequency the majority of the vibration increase occurred (as discussed in Chapter 4).

Fig. 5.3 summarizes the parameters to monitor angular contact bearings and their threshold limits.

Angular contact bearings—parameters to monitor

- Bearing bracket axial vibration—An increase by 20% of baseline should warrant a vibration frequency check to analyze what is causing the vibration increase.
- Bearing bracket temperature—If temperature increases by 20%, look at oil condition and water content in bearing bracket
- Remember to always look at performance on same timeline as axial vibrations, as thrust force will change depending on where you are operating on the performance curve!!!

Fig. 5.3 Parameters to monitor for antifriction angular contact bearings.

HYDRODYNAMIC THRUST BEARINGS

Just like hydrodynamic journal bearings, hydrodynamic thrust bearings use oil to support the rotor. The bearing itself has a clearance (depends on bearing design and equipment manufacturer but usually between 14 and 18 thousandths of an inch total clearance) on the thrust disc, and oil is injected between the bearing and thrust disc. The oil wedge created (the oil between the bearing and thrust disc) is only about 20–25 μm tick, about the thickness of a human hair. Oil can handle a load of approximately 500 psi before it breaks down, so the equipment OEM selects a bearing design and size (and size of the balance drum appropriately) to be loaded at a maximum of 250 psi at normal operating conditions. See Figs. 5.4–5.6 for pictures of a typical hydrodynamic tilting pad thrust bearing which is used in the majority of compressors and steam turbines.

For a hydrodynamic thrust bearing, vibration probes (the same as for a hydrodynamic journal bearing) charged with a DC current measure the axial position of the shaft in mils or mm. A change in axial position by 20% should warrant further investigation into the cause.

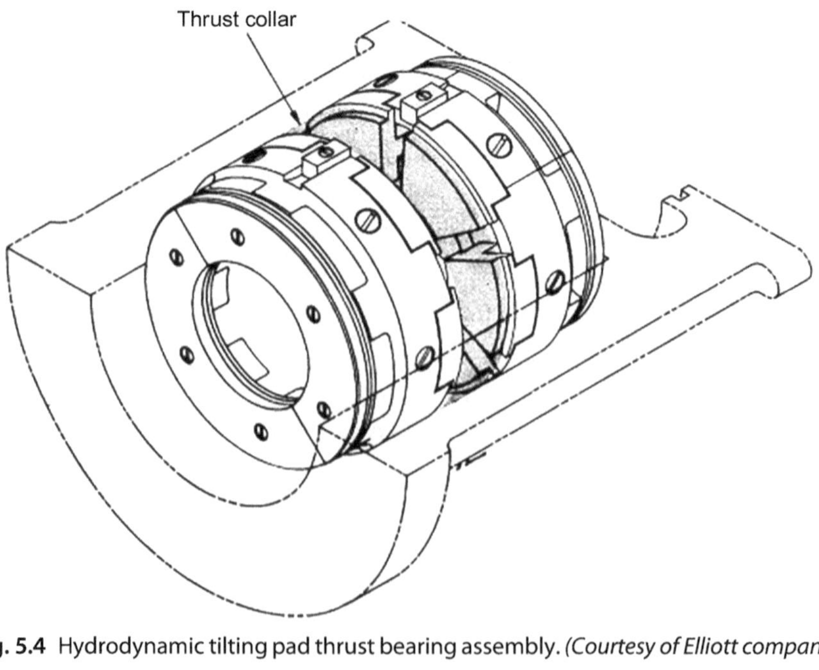

Fig. 5.4 Hydrodynamic tilting pad thrust bearing assembly. *(Courtesy of Elliott company).*

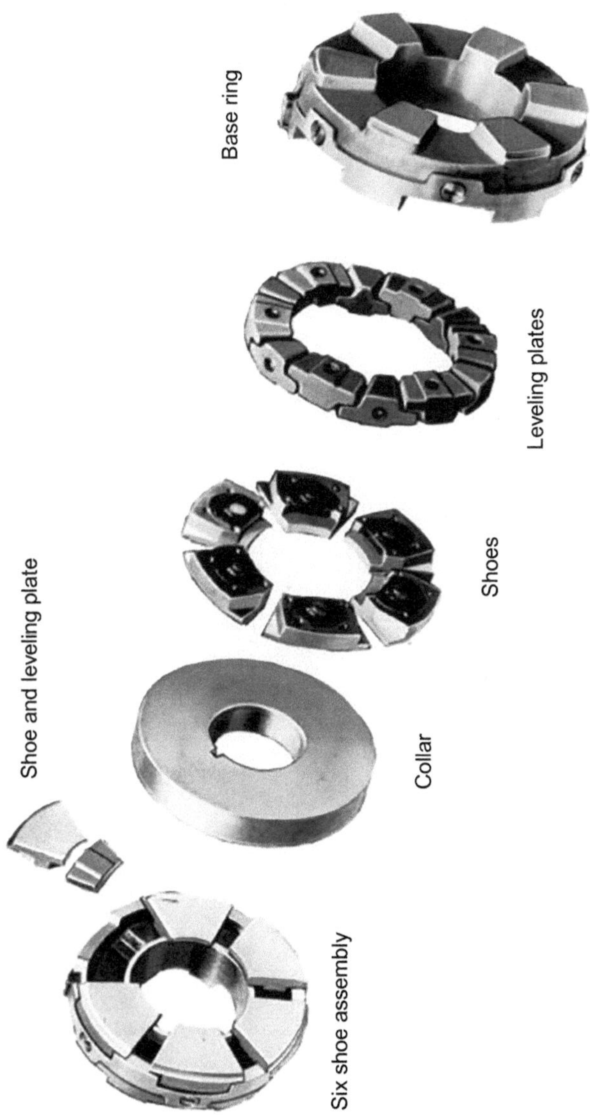

Fig. 5.5 Tilting pad thrust bearing components. *(Courtesy of Kingsbury, Inc).*

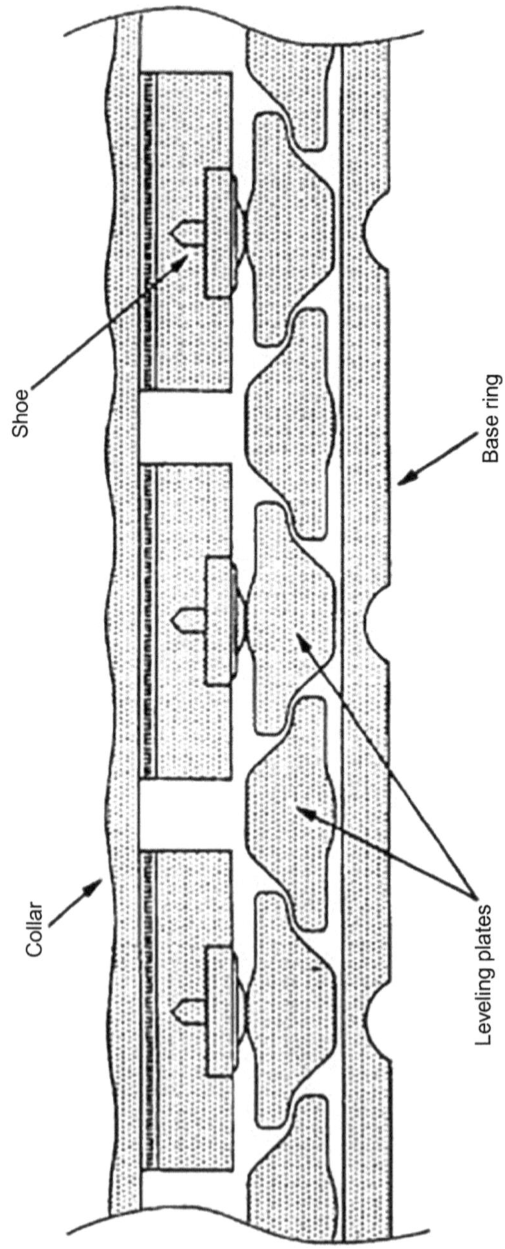

Fig. 5.6 Tilting pad thrust bearing—top view. *(Courtesy of Kingsbury, Inc).*

A number of pads, usually two on each side of the thrust disc, are fitted with RTDs to measure temperature of the pads, which corresponds to load on the thrust bearing. If you see an axial position shift of 20% or more, you should see the pad temperature increasing in the direction of the shaft movement. If not, this indicates that the rotor has not exerted more loads on the thrust pads. If thrust pad temperature has increased by 20%, the cause should be investigated.

The final, and probably most important, parameter to monitor for a hydrodynamic thrust bearing is the balance line differential pressure. See Fig. 5.7 that shows the concept of the balance drum.

The purpose of the balance drum is to limit the load on the thrust bearing, by taking a large differential pressure (many times close to the full differential across the machine) across it. The pressure on one side of the balance drum is total discharge pressure, and the balance drum has a tight clearance labyrinth seal on the out diameter and a port on the other side that goes back many times to suction of the machine to equalize the pressure on the other side very close to suction pressure. This large differential pressure acts on the cross-sectional area of the balance drum to push the rotor in the opposite direction of normal thrust in the machine.

Usually there are pressure taps, one close to the casing where the balance line exits and the other close to where it goes back into the machine (suction

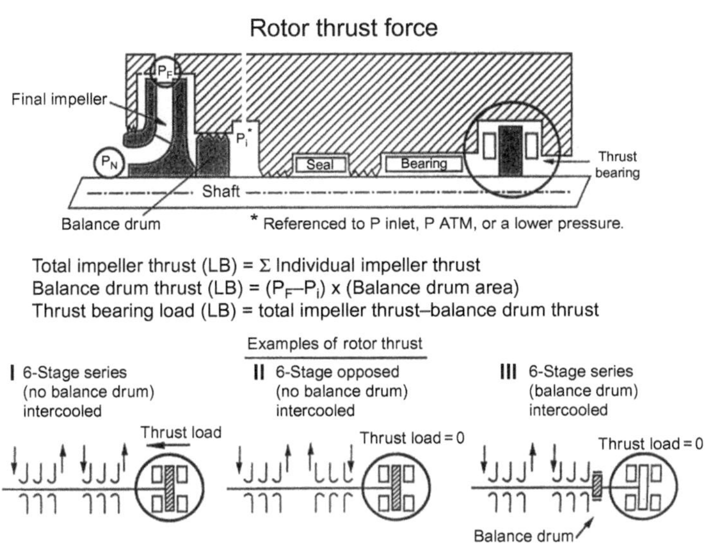

Fig. 5.7 Balance drum.

or a lower pressure stage). The differential pressure is then measured across the balance line. The key, again, is to have a baseline of the balance Line DP and if this DP increases over time, it indicates that the labyrinth seal has worn around the balance drum. If this happens, you will see an increase in axial position toward the active pads and a corresponding increase in pad temperature in that direction. If the balance drum laby is worn, this requires opening the case to fix, while replacing just the thrust bearing will not do anything. Therefore, it is essential to monitor the balance line DP to assure you do not perform any excessive unnecessary maintenance on the thrust bearing. A balance drum laby replacement can be done at the next planned shutdown. It is highly recommended to have a DP transmitter installed in order to have this parameter brought into the DCS and trended in real time.

See Table 5.2 which is a spreadsheet detailing the parameters needed to be monitored and trended for thrust bearings on a steam turbine-driven compressor train.

See Fig. 5.8 for a summary of parameters to monitor for a hydrodynamic thrust bearing and threshold limits.

Table 5.2 Bearing Monitoring Parameters for a Steam Turbine Driven Compressor Train

	Component Condition Monitoring Worksheet
Item #:	
Date/Time:	
Journ. brgs.	
Compressor DE horiz. vibes (micron)	
Compressor DE vert. vibes (micron)	
Compressor DE pad temp (Deg C)	
Compressor DE pad temp (Deg C)	
Compressor NDE horiz. vibes(micron)	
Compressor NDE vert. vibes (micron)	
Compressor NDE pad temp (Deg C)	
Compressor NDE pad temp (Deg C)	
Steam turbine DE horiz. vibes (micron)	
Steam turbine DE vert. vibes (micron)	
Steam turbine DE pad temp. (Deg. C)	
Steam turbine DE pad temp. (Deg. C)	
Steam turbine NDE horiz. vibes (micron)	
Steam turbine NDE vert. vibes (micron)	
Steam turbine NDE pad temp. (Deg. C)	
Steam turbine NDE pad temp. (Deg. C)	

Table 5.2 Bearing Monitoring Parameters for a Steam Turbine Driven Compressor Train—cont'd

	Component Condition Monitoring Worksheet
Thrust brgs.	
Compressor displ.	
Compressor displ.	
Compressor active pad temp. (Deg C)	
Compressor active pad temp. (Deg C)	
Compressor inactive pad temp. (Deg C)	
Compressor inactive pad temp. (Deg C)	
Balance line diff. P (kg/cm^2)	
Steam turbine displ.	
Steam turbine displ.	
Steam turbine active pad temp. (Deg. C)	
Steam turbine active pad temp. (Deg. C)	
Steam turbine inactive pad temp. (Deg. C)	
Steam turbine inactive pad temp. (Deg. C)	

Hydrodynamic thrust bearing—parameters to monitor

- Axial position—A change by 20% should be confirmed by pad temperature increase. Typical alarm limit is approximately 15—20 mils (0.4 to 0.5 mm)
- Pad temperature—An increase in 20% indicates a load increase on thrust bearing. Alarm for thrust pad temperature is around 220 F (108 C)
- Balance line DP—Should be trended, an increase corresponding with increased axial displacement towards the active pads and increased pad temperature indicates the case should be opened to replace the balance drum labyrinth seal at next planned shutdown.

Fig. 5.8 Hydrodynamic thrust bearing parameters to monitor.

Note the comment in Fig. 5.8 regarding axial displacement typical alarm. This is assuming that the thrust bearing monitor is set up where 0 is in the middle of the total thrust bearing clearance. This is the recommended setup so you can easily see that a positive value is toward the suction side pads and a negative number is toward the discharge pads. See Fig. 5.9 showing a typical thrust bearing monitor, illustrating this.

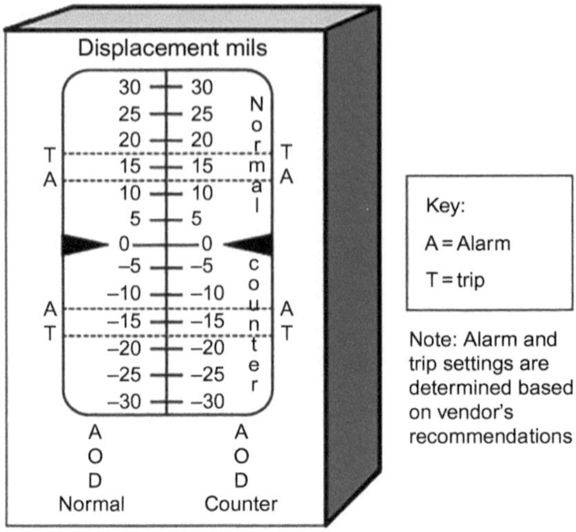

Fig. 5.9 Typical thrust bearing monitor.

Note, the alarm and trip settings shown are recommended based on API, but need to be based on the vendor recommendations for your specific machine.

Now, we will be discussing the key parameters to monitor for component # 4, seals, in all types of rotating machinery.

CHAPTER SIX

Seals

Chapter Outline

We will now focus on the key parameters to monitor seals for all machinery. In this chapter we will discuss pump mechanical seals, compressor oil seals, dry gas seals, and steam seals. Note that some of the parameters discussed will include things to check in the support systems, as it is unavoidable, but there will be more detail on the support systems in Chapter 7.

PUMP MECHANICAL SEALS

Pump mechanical seals minimize the leakage of the process in the form of a vapor across the contacting seal faces. See Fig. 6.1 for a diagram of a typical single pump mechanical seal with components detailed.

The pumped liquid will surround the primary ring (rotating in this case) and try to make its way out between the interface of the primary and mating rings (stationary). The geometry of the faces is designed in conjunction with the operating conditions (pressure) in order to develop a closing force on the two seal faces, and this force breaks down the pressure of the liquid to a point where it becomes a vapor.

See Fig. 6.2 which describes what happens at the seal interface (in between primary and mating rings).

Scenario 1 shows the liquid vaporizing close to the OD of the seal rings. This is not ideal because you will not get optimal lubrication and heat removal at the seal interface. In scenario 1 the seal rings will overheat and be damaged in due time. Scenario 3 outlines a case where the liquid does not vaporize across the seal faces, and this is not good because liquid is incompressible, so the amount of leakage to the atmosphere will be excessive. The scenario in the middle of Fig. 6.2 is what the seal designers aim for.

Forsthoffer's Component Condition Monitoring
https://doi.org/10.1016/B978-0-12-809599-7.00006-3

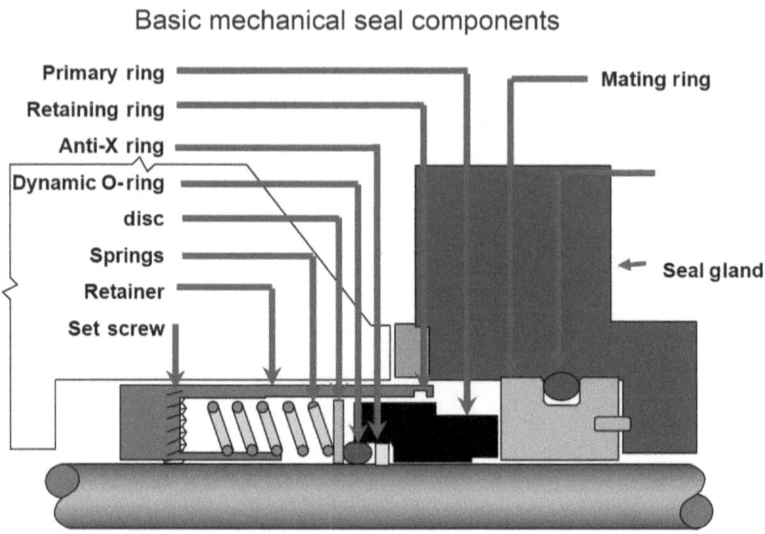

Fig. 6.1 Typical single mechanical seal.

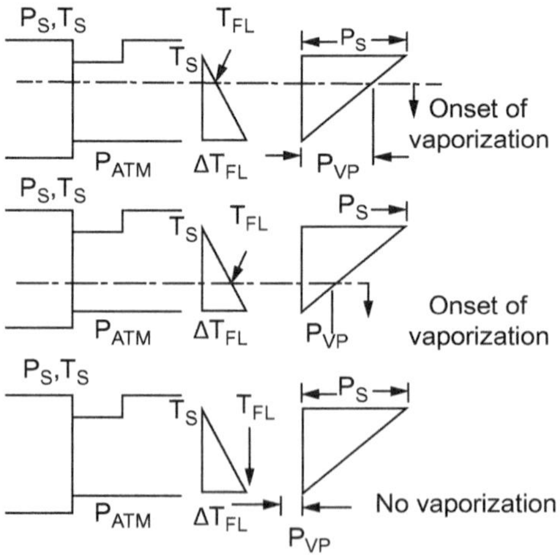

Fig. 6.2 Onset of vaporization at the seal interface.

This shows the liquid vaporizing close to the inner diameter of the seal. It allows for proper cooling of the seal faces, while also converting the liquid to a compressible vapor that can be controlled for minimal leakage.

You can read more about the details of mechanical seal design in a number of books, but the reason it is being discussed here is that the seal designer needs to design the seal geometry such that the liquid is converted to a vapor at a proper point (usually 75% from outer diameter to inner diameter) across a seal face that may be about 3 mm wide. This is precise and is based on the design conditions of the pump, which can change when operating in the field.

Now, looking back at Fig. 6.2 again, what could make your seal go from scenario 2 (ideal) to scenario 1 (inadequate lubrication)?

1. Increase in fluid temperature
2. Decrease in specific gravity of process fluid
3. Decrease in sealing chamber (area around the seal) pressure

Items 1 and 2 both relate to temperature and recall our discussion on centrifugal pump performance back in Chapter 3. We stated that at low flows (typically below 50% of the BEP flow) the efficiency of the pump takes a drastic drop-off resulting in increased process temperature. This is why EROE (as discussed in Chapter 3) is the most important parameter for pump mechanical seal condition. See Fig. 6.3 which describes the concept of EROE again.

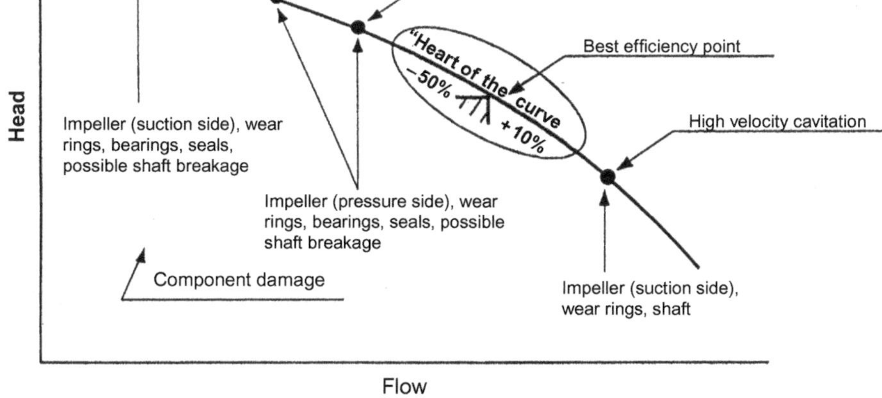

Fig. 6.3 Concept of EROE.

The other key parameters to monitor a pump mechanical seal relate to the fluid properties in the seal chamber and the components within the seal flush system.

In order to accurately monitor the seal chamber pressure, you need to have a pressure gauge installed in the seal chamber. This is not usually available but typically can be fitted in an unused pump seal chamber port (if available) or in the flush system itself. The seal designers want to always maintain the seal chamber pressure at least 50 psi (3.5 bar) above the process fluid's vapor pressure. This will assure the process fluid stays a liquid across the seal faces as much as possible.

See Table 6.1 which identifies all of the parameters to monitor for pump mechanical seal condition.

Fig. 6.4 summarizes the key parameters to monitor for a pump mechanical seal and their threshold limits.

Table 6.1 Pump Mechanical Seal Parameters to Monitor

	Pump Component Condition Monitoring			
Performance				
Item #				
Date				
Input				
P_1 (bara)				
P_2 (bara)				
S.G.				
Pump speed (RPM)				
Flow rate (m^3/h)				
Calculate				
Head (m)	–	–	–	–
Flow determination[a]				
Control valve position				
CV for valve and trim type[b]				
Fluid S.G.				
Control valve ΔP				
Calculated valve flow	–	–	–	–
Motor amps				
Volts				
Power factor				
Motor Eff'y				
Calculated power	–	–	–	–
Flow from pump curve[c]				
T1 (Deg. C)				
T2 (Deg. C)				

Table 6.1 Pump Mechanical Seal Parameters to Monitor—cont'd

	Pump Component Condition Monitoring			
CP (specific heat)				
Calculated head	–	–	–	–
Calculated pump Eff'y	–	–	–	–
Flow from pump curve[c]				
Pump maintenance required[d]				
EROE determination				
BEP flow[e]				
EROE min flow	–	–	–	–
EROE max flow	–	–	–	–
Is pump in EROE	–	–	–	–
EROE targets for operations				
Flow				
Amps				
Pump Δ T (measured on inlet and discharge pipes)				
	Pump Mechanical Seal and Flush System Condition Monitoring			
Mechanical seal				
Item #				
Group name				
Seal chamber condition				
Pressure (bara)				
Temperature (deg C)				
Fluid sample?				
Sample results				
Is pressure as designed?				
Is temp. as designed?				
Action required				
Seal flush system condition				
Temp difference across orifice (deg C)				
Is orifice plugged?				
Temp difference across cyclone				
Is cyclone plugged?				
Temp difference across cooler				
Is cooler fouled?				
External flow (plan 32) as designed?				
Plan 52/53 res. diff temp. (deg C)				
Reservoir level				

Continued

Table 6.1 Pump Mechanical Seal Parameters to Monitor—cont'd

	Pump Component Condition Monitoring
Reservoir pressure (bara)	
Vented to flare?	
Blocked in?	
Barrier fluid sample?	
Steam condition if steam quench installed?	
Action required	
Overall required action	

[a]If a flow meter is not available for the pump flow will be determined by one of the alternative methods noted in Table 7 of the supplementary manual. These alternative methods are: 1. Portable Ultrasonic Flow meter, 2. Control Valve Position to calculate flow, 3. Motor Amps to calculate Power, 4. Pump Δ T to calculate pump efficiency.
[b]The CV can be found on the Valve Manufacturers curve for specific valve type and trim. Note that most of the major valve manufacturers post these curves on their websites.
[c]Flow will be estimated using the original shop test curve. Note that if the pump is not in good condition these estimates will not be very accurate.
[e]Flow at Highest Efficiency taken from the shop test curve in the supplementary manual.
[d]Pump Maintenance should be considered if the head and flow (operating point) when plotted on the test curve is approx. 10% below test curve flow for calculated head.

Pump mechanical seals—parameters to monitor

- **EROE—Assure your pump is operating within EROE for Optimal Seal Reliability**
- **Seal chamber pressure—Pressure needs to be >50 psi (3.5 bar) above the process fluid vapor pressure at its operating temp.**
- **Seal chamber temperature—If temperature increases by 10%, assure seal chamber pressure is still adequate as per above.**
- **Fluid specific gravity—If fluid can change properties or if temperature increases significantly, take a sample and check S. G.**
- **Remember to check all flush system components to assure flow of flush liquid is being provided to the seal!!!**

Fig. 6.4 Parameters to monitor for pump mechanical seals.

As stated in Fig. 6.4, it is essential to check all components within your specific seal flush systems to assure there is no blockage and that you are circulating proper cool, clean flush to the seals.

COMPRESSOR OIL SEALS

There are a number of different designs of oil seals for compressor, all of which essentially have the same key parameters to monitor for condition.

Figs. 6.5–6.7 show a few different types of oil seals found in compressors.

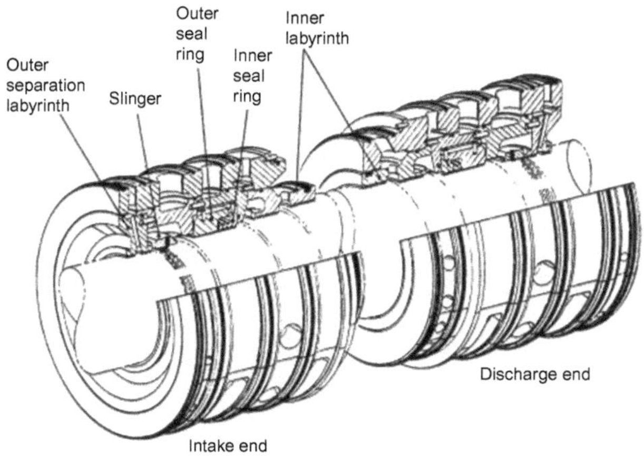

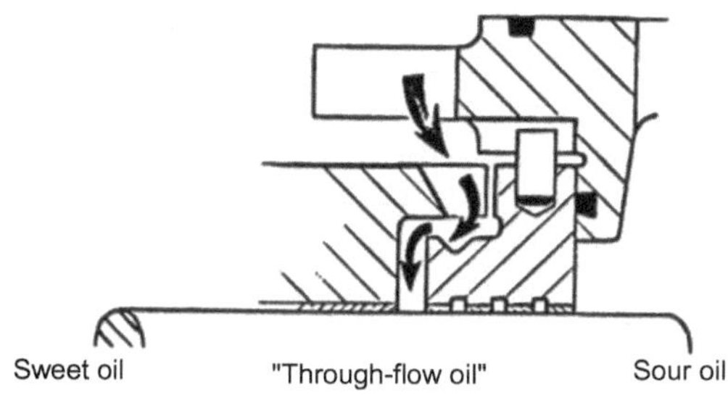

Fig. 6.5 Bushing seal.

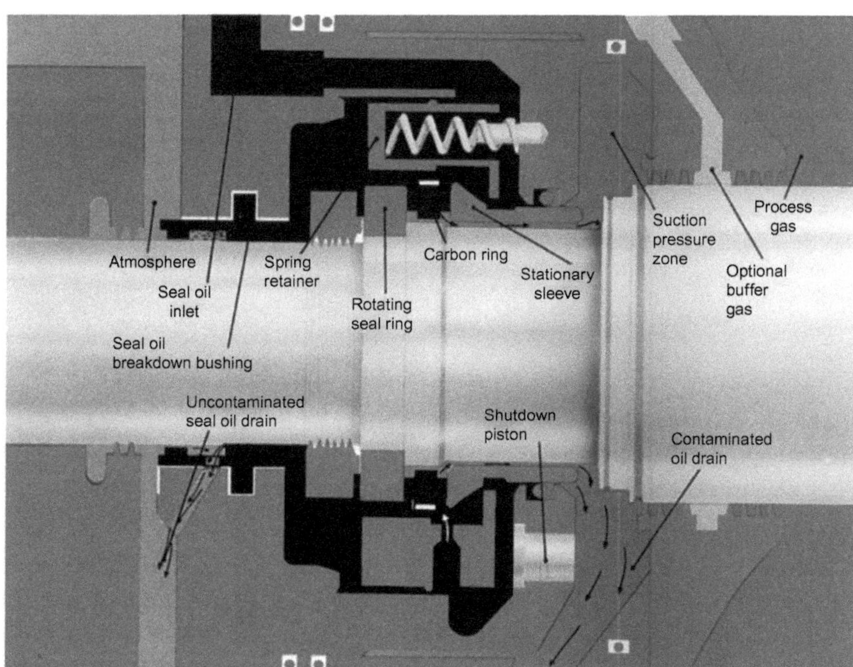

Fig. 6.6 Elliott Iso-carbo seal. *(Courtesy of Elliott Co).*

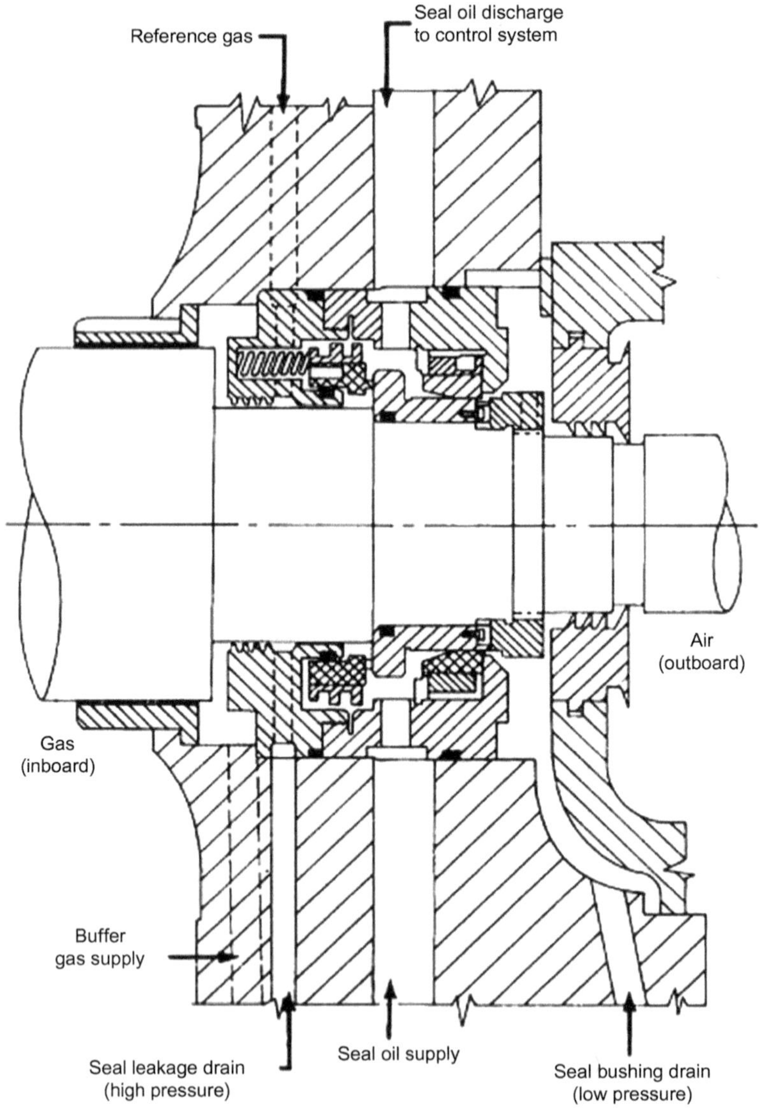

Fig. 6.7 Contacting oil seal. *(Courtesy of IMO Industries.)*

All of these seals have a seal oil inlet at a differential pressure above the reference gas pressure; however the bushing seal (Fig. 6.5) uses a level control valve and an overhead seal oil tank to maintain a DP of approximately 5 psid (0.3 bar) and the seals in Figs. 6.6 and 6.7 utilize a DP control valve to maintain the differential pressure. This is because they are essentially both contacting type seals and can handle a higher DP. In all compressor oil seals, the leakage across the process seal combines with the process or buffer gas

Compressor oil seals—parameters to monitor

- **Sour oil leakage—Should be checked monthly. If leakage increases by 20%, the oil system components need to be checked.**
- **Seal oil drainer level—Zero level could mean the LCV is hung open, allowing gas to get back to the reservoir. Full drainer could mean the oil is entering the compressor case.**
- **Sniff flanges for gas on the atmospheric drain to assure gas is not present.**

Fig. 6.8 Parameters to monitor for compressor oil seals.

and goes to sour oil drainers at a normal rate of anywhere from 5 to 40 gal per day. The drainers separate some of the gas from the contaminated oil and return the oil to a degassing tank before entering the seal oil reservoir (usually the same as the lube oil reservoir). Leakage across the atmospheric side seal flows directly back to the reservoir, as it is not contaminated.

The key parameters to monitor include seal oil control valve position, sour oil leakage, and drainer level.

If level in the sour oil drainer is not present, you should immediately sniff the reservoir and take an oil sample to check for gas in the oil, as this could be a serious safety hazard. A spreadsheet will be available in Chapter 7 containing all the parameters to check on the Seal oil system. See Fig. 6.8 for a summary of parameters to check for a compressor oil seal.

DRY GAS SEALS

Dry gas seals operate very similar to a contacting oil seal and pump mechanical seal, except they have a small (less than 6 μm thick) gas film that separates the seal faces from each other and minimizes leakage to the atmosphere.

Figs. 6.9 and 6.10 shows two typical tandem gas seal configurations that are used in most compressor applications.

The seal arrangement at the top of Fig. 6.9 uses a clean and dry primary seal gas supply controlled by a DP control valve over reference gas pressure (just like an oil seal) and the majority (over 99%) enters the process. The small amount that does leak across the primary seal (up to a few SCFM) will mostly exit through the primary vent to flare with the remainder leaking across the secondary seal to the secondary vent. A separation seal (can be a labyrinth, abradable laby, or carbon bushings) is located between the seal

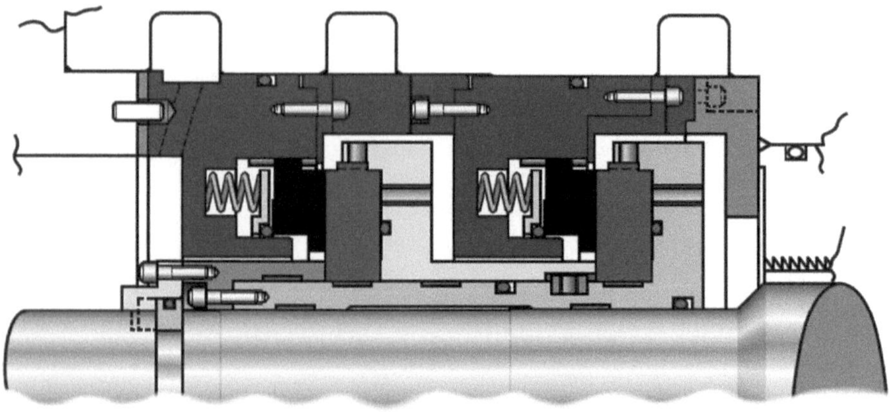

Fig. 6.9 Tandem dry gas seal without intermediate laby.

and bearings, and N2 is injected at a pressure higher than the secondary vent in order to keep process gas from enter the bearing. The nitrogen then exits with the process gas leakage from the secondary seal through the secondary vent. The nitrogen injected into the separation seal is also supposed to keep lube oil from entering the dry gas seal. If enough oil enters the seal area it can build up in the secondary seal drain area and eventually fill up the grooves of the dry gas seal. If the grooves fill up with liquid or contaminants, the seal loses its capability to separate and the faces contact and fail very soon after.

The seal in Fig. 6.10 has only one difference. There is nitrogen (called intermediate nitrogen) supplied between the primary and secondary seals and a labyrinth seal in between where the nitrogen flows across (from right

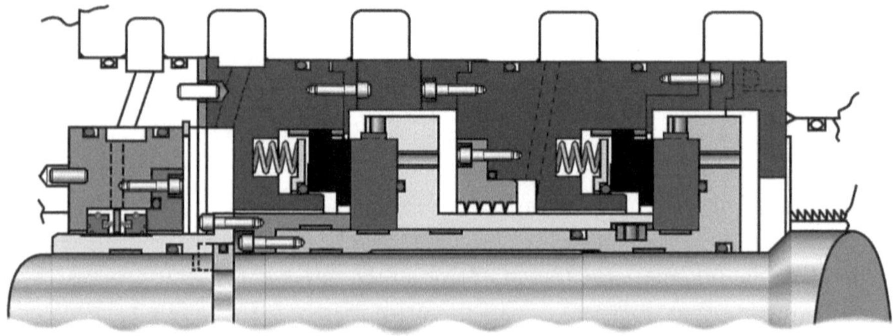

Fig. 6.10 Tandem dry gas seal with intermediate laby.

to left in the picture) and goes out through the primary vent with the primary seal leakage. The advantage of this seal is that the secondary seal is in a nitrogen-rich environment, preventing process gas from leaking to the atmosphere. It is highly recommended to have the intermediate nitrogen controlled at a differential pressure above primary vent in order for this to positively prevent process gas from entering the secondary seal cavity.

Key parameters to monitor for a dry gas seal are as follows:

1. Primary seal gas differential pressure—monitor and trend control valve position
2. Primary vent flow or pressure—high for primary seal leakage, low for secondary seal leakage
3. Intermediate N2 DP (If available)
4. Primary vent control valve position (if available)—if opening, the primary seal leakage has increased, if closing the secondary seal leakage has increased.
5. Separation N2 pressure or differential pressure (if equipped with DP control)

Tables 6.2 and 6.3 are spreadsheets showing what to typically monitor on the two types of dry gas seals illustrated in Figs. 6.9 and 6.10.

Notable from those tables, the two most important parameters that are not always monitored and trended are primary vent control valve position and secondary drain oil contamination. Most newer dry gas seals have a set back pressure in the primary vent to optimize the life of the secondary seal by

Table 6.2 Dry Gas Seal Monitoring without Intermediate N2

DGS System Monitoring

Equipment #: **Item**	**Observations**	**Date:** **Time:** **Comments**
Primary gas filter DP (Psid)		
Primary gas supply DP		
Prim. vent pressure suct. (PSIG)		
Prim. vent pressure Disch. (PSIG)		
Primary vent back pressure (PSIG)		
Primary vent to SV DP suction (PSID)		
Primary vent to SV DP discharge (PSID)		
Primary vent control valve position (%)		
Separation gas filter DP (PSID)		
Sep gas to SV DP (PSID)		

Table 6.3 Dry Gas Seal Monitoring with Intermediate N2

DGS System Monitoring

Equipment #: Item	Observations	Date: Time: Comments
Primary gas filter DP (kg/cm^2)		
Primary gas supply DP (kg/cm^2)		
Prim. vent flow DE (Nm^3/h)		
Prim. vent flow NDE (Nm^3/h)		
Primary vent control valve position (%)		
Sec. gas filter DP (kg/cm^2)		
Sec. gas supply press. (kg/cm^2)		
Sec. supply flow DE (Nm^3/h)		
Sec. supply flow NDE (Nm^3/h)		
Sec. vent DE flow (Nm^3/h)		
Sec. vent NDE flow (Nm^3/h)		
Oil in sec. drain DE?		
Oil in sec. drain NDE?		

providing a DP across it, but most importantly to monitor leakage of the secondary seal if that pressure drops. The back pressure can be set by a typical pneumatic control valve with electronic positioner. The position of this valve tells you ahead of time if one of the seals is starting to leak excessively.

Regarding the secondary drain, as discussed earlier in this section, oil accumulation should not be allowed to build up in this area. It is highly recommended to drain any oil from this area (using proper PPE) on a regular basis. A sight glass can be installed in this drain to allow for easier observation of oil in this area of the seal.

STEAM SEALS

Any plant with special purpose steam turbine drivers has probably had a time where excessive amounts of water entered the oil system (sound familiar?). The usual source of this water is from steam leakage across the seals that enter the bearing housings and condenses. Therefore, it is essential to assure that steam seal system is working properly and pulling the steam down to the seal gland condenser system.

Figs. 6.11 and 6.12 are diagrams of typical steam turbine labyrinth seals and the steam seal system.

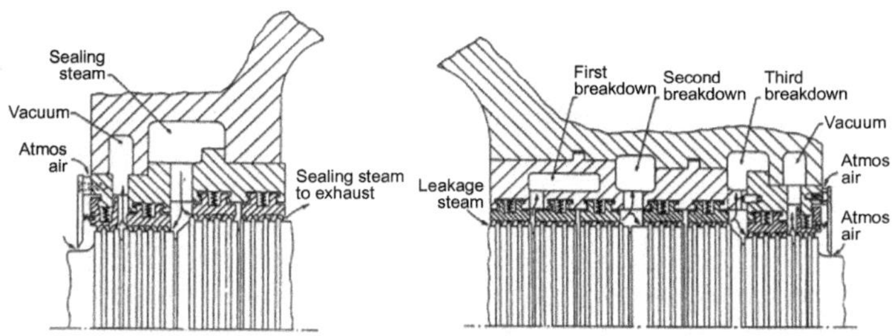

Function: The labyrinth type shaft end seals safely direct the expansion vapor to a defined location and draw in a buffer gas (air) to prevent expansion vapor contact with the bearings.

Fig. 6.11 Typical steam seal schematic.

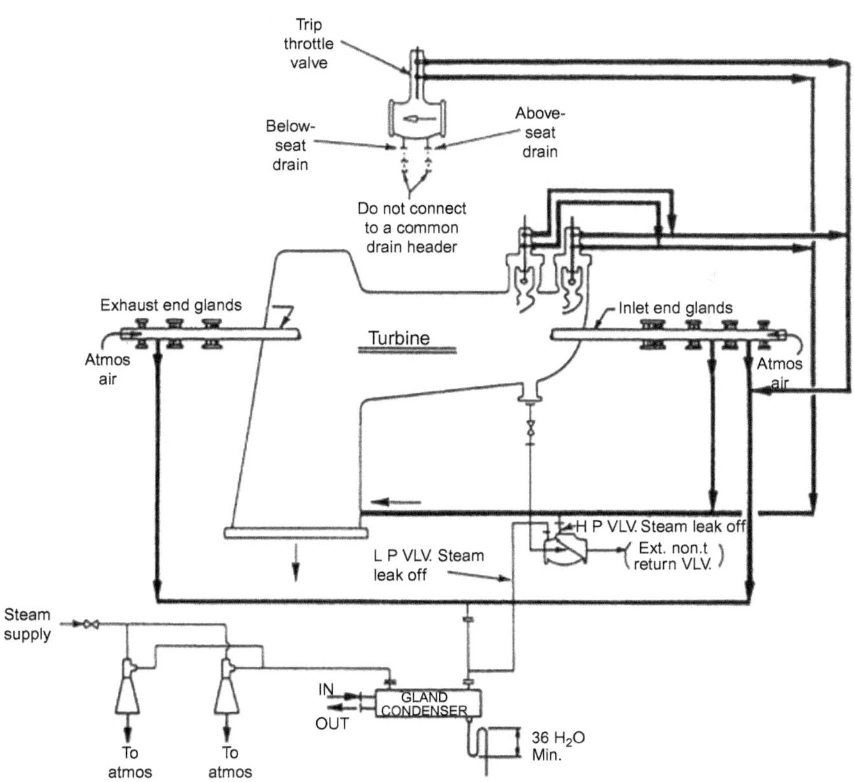

Fig. 6.12 Typical steam seal system.

The steam seals are a series of labyrinths that break down the steam pressure and are connected at one of the ports to a condenser and eductor that utilizes higher pressure steam to pull a vacuum and bring the low pressure steam from the seals down to the condenser. As the eductor nozzles wear over time (from erosion) they lose the ability to pull the steam leakage down to the condenser. So, the key parameter to monitor and trend for the steam seals condition is the pressure in the gland condenser. The pressure should always be a vacuum, if it goes above atmospheric pressure, the steam will not make it to the condenser. Typically, a pressure gauge is fitted on or near the condenser to monitor; however it is highly recommended to have a transmitter installed so this value can be trended. Also, highly recommend, and depicted in Fig. 6.11, is installing two eductors in this system, where you can change over to the other online when the condenser pressure is approaching atmospheric pressure.

In Chapter 7 we will be getting into auxiliary systems for rotating machinery to discuss the key parameters to monitor. Note that the reliability of the components is only as good as the reliability of their support systems.

CHAPTER SEVEN

Auxiliary Systems

Chapter Outline

This chapter will discuss the parameters to monitor for auxiliary systems. Since the parameters of some systems were discussed in Chapter 6 (pump seal systems, dry gas seal systems, and steam seal systems) we will not repeat them here. Rather we will focus on lube oil, seal oil, and gas turbine auxiliary systems).

COMPRESSOR LUBE OIL SYSTEMS

Figs. 7.1 and 7.2 show a couple of different views of a compressor lube oil console, and a typical lube oil system P&ID is captured in Fig. 7.3.

The purpose of a lube oil system is to continuously provide cool, clean oil to the bearings and gear (if applicable) at the proper pressure, temperature, and flow rate. Therefore, the major components of any lube oil system are as follows:

- Reservoir
- Main and auxiliary pumps
- Back pressure control valve (to bypass excessive flow back to the reservoir if PD pumps are selected)
- Coolers—with temperature control valve to control temperature being supplied to bearings.
- Filters—with differential pressure monitoring
- Lube oil supply pressure control valve

Table 7.1 is a spreadsheet indicating the key parameters to monitor on a typical lube oil system.

All the parameters listed in Table 7.1 should be checked initially as a baseline, then re-checked when there is a bearing or gear (if applicable)

Forsthoffer's Component Condition Monitoring
https://doi.org/10.1016/B978-0-12-809599-7.00007-5

Fig. 7.1 Pump side view of lube oil console.

Fig. 7.2 Filter side view of compressor lube oil system.

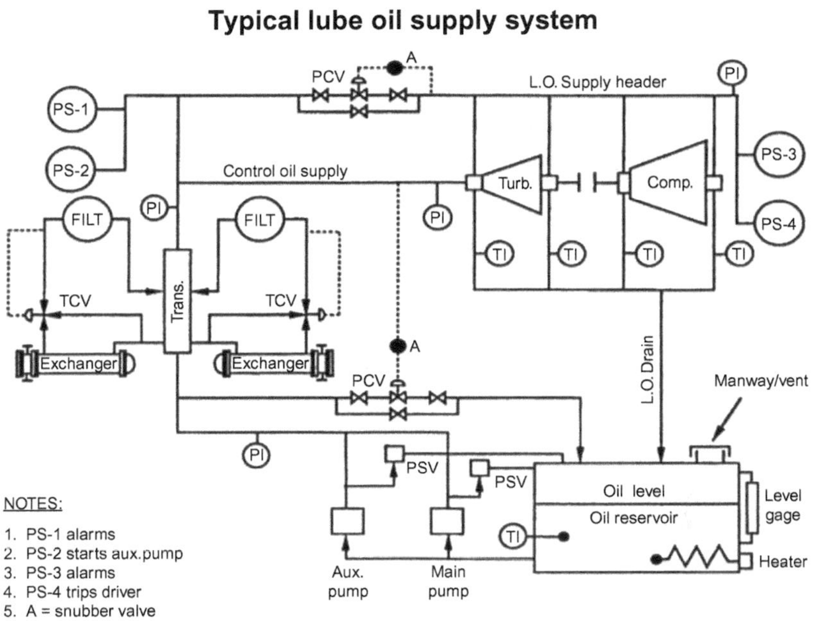

Fig. 7.3 Lube oil system schematic. *(Courtesy of M.E. Crane Consultant).*

Table 7.1 Lube Oil System Condition Monitoring

	Equip #:		**Date**
	System Name:		**Time**
Component/item	**Specified Value**	**Actual Value**	**Comments**
Oil reservoir			
Level			
Oil temp. (C)			
Air in oil?(Y/N)			
Gas in oil?			
Oil sample?			
Pumps			
Aux. pump operating?			
P2 (bar)			
Suction noise?			
Suction filter Δ P (bard)			
Vibration (μm)			
Brg. bracket temp. (C)			

Continued

Table 7.1 Lube Oil System Condition Monitoring—cont'd

	Equip #:		**Date**
	System Name:		**Time**
Component/item	**Specified Value**	**Actual Value**	**Comments**
Couplings			
Noise?			
Strobe findings			
Turbine driver			
Operating speed (RPM)			
Trip speed setpoint (RPM)			
Vibration (μm)			
Brg. bracket temp. (C)			
Gov. hunting?			
Trip lever condition			
Gov. oil condition			
Motor driver			
Operating?			
Vibration (μm)			
Brg. bracket temp. (C)			
Axial shaft movement (μm)			
Fan noise?			
Relief valves			
Passing?			
Set pressure (bar)			
Pump P2 press. (bar)			
Check valves			
Aux. pump turning backward?			
Noise?			
Back pressure valve			
% Open			
Stable?			
Valve noise?			
Set pressure (kPa)			
Maintained pressure (kPa)			
Transfer valves			
One bank operating?			
Noise?			
Coolers			
Δ T oil			
CW valve pos.			
Cooler operating?			
Vent valves open?			

Table 7.1 Lube Oil System Condition Monitoring—cont'd

	Equip #:		**Date**
	System Name:		**Time**
Component/item	**Specified Value**	**Actual Value**	**Comments**
TCVs			
% Open			
Set temp. (C)			
Stable?			
Actual temp. (C)			
Filters			
Δ P (bar)			
Vent valves open?			
Last filter change			
Accumulators			
Precharged pressure (bar)			
Last PM date			
Lube oil PCV			
% Open			
Set pressure (bar)			
Actual pressure (bar)			
Stable?			
Control oil PCV			
% Open			
Set pressure (bar)			
Actual pressure (bar)			
Stable?			
Lube oil rundown tank (or emerg. pump)			
Pump or tank?			
Pump operating?			
Tank overflow			
Lube oil supply lines			
Leaks?			
Noise?			
Vibration (μm)			

condition change. Of the parameters listed in the table it is worth noting a few key items that may not be normally checked. They are as follows:

- Back pressure control valve position (only applicable with PD pumps)—This is the most important parameter to monitor any oil system as it essentially tells the flow of the pumps. If the valve begins closing, it indicates that either the pumps are wearing or flow is being bypassed

somewhere else in the system (relief valves passing). As soon as this valve begins to close, the cause should be investigated, since once the valve is fully closed it is only a matter of time before the system pressure drops enough for the auxiliary pump to hopefully start up.
- Temperature control valve position—This indicates the effectiveness of the coolers. As the control valve bypasses less flow around the coolers and when the setpoint cannot be met, the cooler should be switched over to the nonoperating cooler. If the temperature setpoint can still not be met, it indicates the cooling water temperature or flow is not proper.
- Filter differential pressure—This should be used to change over to the nonoperating filter at a set value (usually 15 psi or 1 bar), and the dates of filter changeover should be documented. If the frequency of filter changeover has increased, this indicates that there is entrained dirt/buildup in the system and a system flush should be performed during the next planned shutdown.
- Accumulator precharge check—This is a sensitive check that many facilities do not want to perform online, since you need to drain the accumulator to check the N2 precharge. That said, if your accumulator precharge is not present, the protection is already gone, and your compressor train is at risk. This is recommended to be performed online every three months. On new systems (or systems upgrading to an accumulator) it is highly recommended to install two 100% capacity accumulators, so when one is checked online, you are fully protected against any transients.

Another important component to be checked regularly on all systems is the condition of the oil. Fig. 7.4 discusses oil analysis checks and the threshold limits for them.

Oil analysis checks

- Two samples should be taken, one at a low point on the reservoir and another downstream of the filters for the following:
  - Viscosity—should be within the oil specification
  - Water content—should never be greater than 200 ppm. If so, install a vacuum dehydration unit as soon as possible and determine the cause of water entering the system.
  - Flash point—A flash point less than 200 C indicates there may be gas in the system. If this is the case, rent or purchase a vacuum degassing unit as soon as possible.
  - Particle count—should be checked downstream of the filters to confirm if particles larger than the filter element mesh are passing.

Fig. 7.4 Oil analysis checks.

SEAL OIL SYSTEMS

The purpose of the seal oil system is very similar to the lube oil system and many times they are a combined system. A seal oil system continuously provides cool, clean seal oil to the seals at the proper differential pressure over seal gas reference pressure, temperature, and flow rate. It does this with the same components as in a lube oil system with the addition of a seal oil supply valve (either level control or differential pressure control), a seal oil overhead tank (on floating oil bushing seals), sour oil drainers, and a degassing tank.

Table 7.2 is a spreadsheet detailing all of the components to monitor in a typical combined lube and seal oil system.

Table 7.2 Lube and Seal Oil Systems, Parameters to Monitor

	Equip #:	**Date**
	System Name:	**Time**
Component/item		**Comments**
Combined lube system		
Date		
Oil reservoir		
Level		
Oil temp.(C)		
Air in oil?(Y/N)		
Gas in oil?		
Oil sample?		
Pumps		
Aux. pump operating?		
P2 (barg)		
Suction noise?		
Suction strainer Δ P (kPa)		
Vibration (μm)		
Brg. bracket temp. (C)		
Couplings		
Noise?		
Strobe findings		

Continued

Table 7.2 Lube and Seal Oil Systems, Parameters to Monitor—cont'd

Component/item	Equip #: / System Name:	Date / Time / Comments
Turbine driver		
Operating speed (RPM)		
Trip speed setpoint (RPM)		
Vibration (μm)		
Brg. bracket temp. (C)		
Gov. hunting?		
Trip lever condition		
Gov. oil condition		
Motor driver		
Operating?		
Vibration (μm)		
Brg. bracket temp. (C)		
Axial shaft movement (μm)		
Fan noise?		
Relief valves		
Passing?		
Set pressure (kPa)		
Pump P2 press. (kPa)		
Check valves		
Aux. pump turning backward?		
Noise?		
Back pressure valve		
% Open		
Stable?		
Valve noise?		
Set pressure (kPa)		
Maintained pressure (kPa)		
Transfer valves		
One bank operating?		
Noise?		
Coolers		
Δ T oil		
CW valve pos.		
Cooler operating?		
Vent valves open?		
TCVs		
% Open		
Set temp. (C)		
Stable?		
Actual temp. (C)		

Table 7.2 Lube and Seal Oil Systems, Parameters to Monitor—cont'd

Component/item	Equip #: / System Name:	Date / Time / Comments
Filters		
Δ P (kPa)		
Vent valves open?		
Last filter change		
Accumulators		
Precharged pressure (kPa)		
Last PM date		
Lube oil PCV		
% Open		
Set Pressure (kPa)		
Actual pressure (kPa)		
Stable?		
Control oil PCV		
% Open		
Set pressure (kPa)		
Acual pressure (kPa)		
Stable?		
Lube oil rundown tank (or emerg. pump)		
Pump or tank?		
Pump operating?		
Tank overflow		
Lube oil supply lines		
Leaks?		
Noise?		
Vibration (μm)		
Seal oil supply valve		
Position		
Stable?		
Seal oil supply pressure		
Seal oil OH tank		
Level		
S.O. differential pressure		
Drainers		
Sour oil leakage rate		
Level		
Flow through vent orifice?		

Continued

Table 7.2 Lube and Seal Oil Systems, Parameters to Monitor—cont'd

Component/item	Equip #: / System Name:	Date / Time / Comments
Separate Seal oil system		
Date		
Oil reservoir		
Level		
Oil temp.(C)		
Air in oil?(Y/N)		
Gas in oil?		
Oil sample?		
Pumps		
Aux. pump operating?		
P2 (barg)		
Suction noise?		
Suction strainer Δ P (kPa)		
Vibration (μm)		
Brg. bracket temp. (C)		
Couplings		
Noise?		
Strobe findings		
Turbine driver		
Operating speed (RPM)		
Trip speed setpoint (RPM)		
Vibration (μm)		
Brg. bracket temp. (C)		
Gov. hunting?		
Trip lever condition		
Gov. oil condition		
Motor driver		
Operating?		
Vibration (μm)		
Brg. bracket temp. (C)		
Axial shaft movement (μm)		
Fan noise?		
Relief valves		
Passing?		
Set pressure (kPa)		
Pump P2 press. (kPa)		
Check valves		
Aux. pump turning backward?		
Noise?		

Table 7.2 Lube and Seal Oil Systems, Parameters to Monitor—cont'd

	Equip #:	Date
	System Name:	Time
Component/item		Comments
Back pressure valve		
% Open		
Stable?		
Valve noise?		
Set pressure (kPa)		
Maintained pressure (kPa)		
Transfer valves		
One bank operating?		
Noise?		
Coolers		
Δ T oil		
CW valve pos.		
Cooler operating?		
Vent valves open?		
TCVs		
% Open		
Set temp. (C)		
Stable?		
Actual temp. (C)		
Filters		
Δ P (kPa)		
Vent valves open?		
Last filter change		
Accumulators		
Precharged pressure (kPa)		
Last PM date		
Seal oil supply valve		
Position		
Stable?		
Seal oil supply pressure		
Seal oil OH tank		
Level		
S.O. differential pressure		
Drainers		
Level		
Flow through vent orifice?		

GAS TURBINE AUXILIARY SYSTEMS

Besides a lube oil system, as outlined earlier in this chapter, gas turbines have numerous other auxiliary systems due to the high temperatures they operate at and how susceptible they are to fouling. Figs. 7.5 through 7.7 highlight the auxiliary systems that apply to gas turbines.

Table 7.3 details the components to monitor within the auxiliary systems for gas turbines.

If any parameter changes by 10% or more, the OEM should be consulted and the data should be analyzed to determine when maintenance should be performed at a convenient time.

Now that we have all the tools, we will now discuss the importance of using these data to minimize PMs and the frequency of the required PMs.

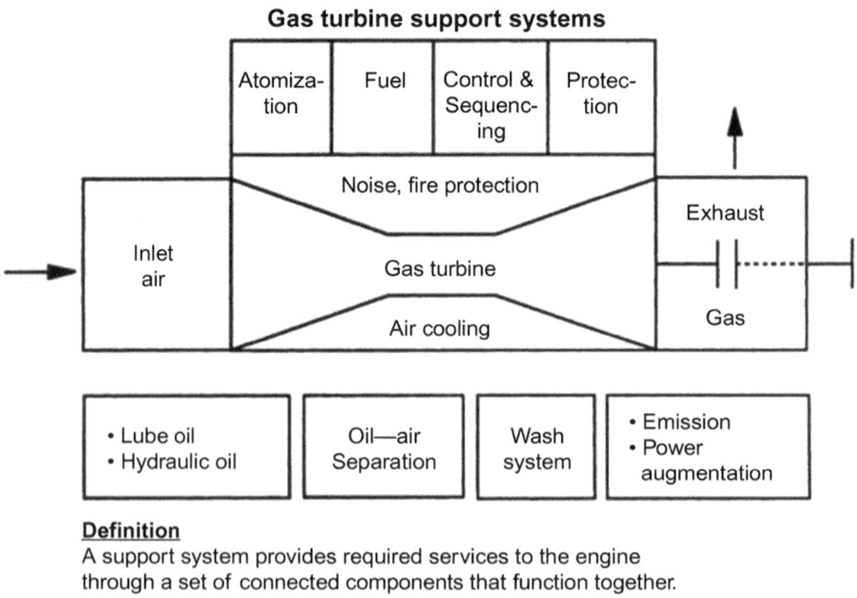

Fig. 7.5 Gas turbine support system overview.

Types of gas turbine support systems

Air/gas handling

- Inlet
  - Filter
  - Silencer
  - Duct
- Exhaust
  - Duct
  - Silencer
  - Heat recovery (HRSG)

Lube/control oil

- Engine lube
- Auxiliary lube
- Driven equipment lube
- Engine control actuation
  - Guide vane
  - Stator vane
  - Trip
  - Control valve
- Air/oil separation
- Scavenging

Control

- Engine start/stop sequencing
- Condition monitoring
- Governor system
- Protection system
- fuel system
  - Gas
  - Liquid
  - Dual
- Atomizing air

Cooling

- Engine internal
  - Industrial
  - Aero-derivative
- Engine external
  - Industrial
- Auxiliary cooling
  - Lube oil coolers (etc.)

Injection

- Emission control
  - H_2O
  - Steam
- Power augmentation
  - H_2O
  - Steam

Engine cleaning

Liquid

- Crank
- On-line

Solid

- On-line (industrial)

Engine enclosures

- Noise
- Pressurization
- Fire protection
  - Halon
  - CO_2

Fig. 7.6 Types of gas turbine support systems.

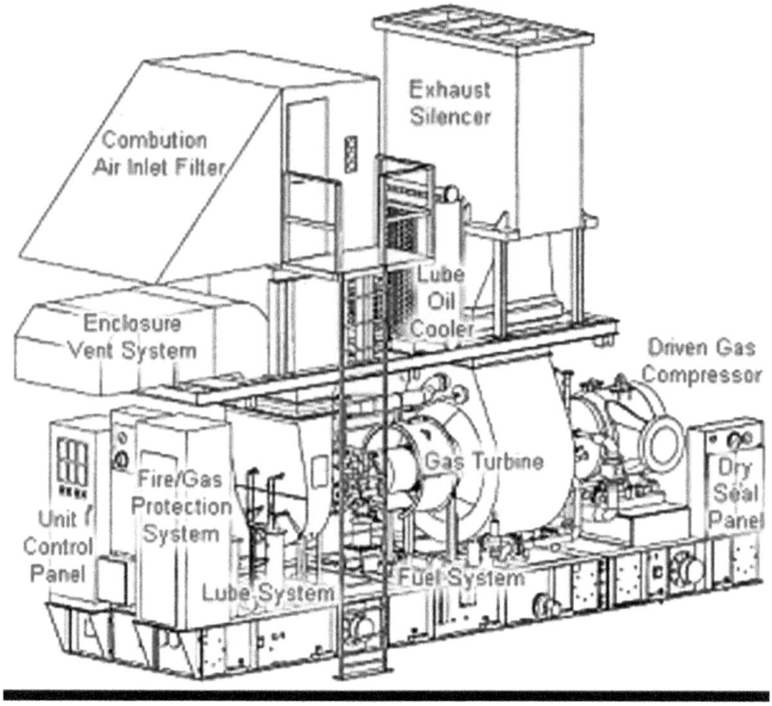

Fig. 7.7 Gas turbine support system schematic.

Table 7.3 GT Aux System Condition Monitoring

	Gas Turbine Component Condition Monitoring
Equip. #	
Date	
Air filtration Sys. condition	
Filter DP (bard)	
Action of self-cleaning air system (active or inactive)?	
Action?	
Variable inlet guide vanes	
Hyd. supply differential press. (bard)	
Guide vane position	
Air flow set point	
Measured air flow	
IGV exhaust temp. reference	
Measured exhaust temp. (C)	
Action?	
Cooling and sealing air sys.	

Table 7.3 GT Aux System Condition Monitoring—cont'd

	Gas Turbine Component Condition Monitoring
Cooling air pressure (bara)	
Other cooling sys. observations	
Air sealing system supply pressure (bara)	
Fuel gas system	
Ambient temp. (C)	
Relative humidity	
Fuel supply pressure (bara)	
Fuel flow rate (kg/hr)	
Fuel pressure upstream of shut off valve (bara)	
Fuel shut off valve position	
Fuel pressure upstream of fuel control valve (bara)	
Fuel control valve position	
Fuel supply pressure to combustors (bara)	
Action?	
Combustion monitoring system	
Ambient temp. (C)	
Relative humidity	
Exhaust temp. 1 (C)	
Exhaust temp. 2 (C)	
Exhaust temp. 3 (C)	
Exhaust temp. 4 (C)	
Exhaust temp. 5 (C)	
Exhaust temp. 6 (C)	
Exhaust temp. 7 (C)	
Exhaust temp. 8 (C)	
Exhaust temp. 9 (C)	
Exhaust temp. 10 (C)	
Exhaust temp. 11 (C)	
Exhaust temp. 12 (C)	
Allowable spread (C)	
Temp spread 1 (C)	
Temp spread 2 (C)	
Temp spread 3 (C)	
Exhaust thermocouple alarm on?	
Combustion alarm on?	
High exhaust temp. spread trip?	
Monitor enable activated?	
Action?	
Exhaust temperature control system	
Compressor Disch. P (bara)	
Fuel stroke reference	
Ambient temperature (C)	
Relative humidity	
Calculated firing temp. (C)	
Action?	

CHAPTER EIGHT

Trust the Method

The whole book has focused on what component condition monitoring is and the parameters you need to monitor for all rotating machinery, but the question is how do we use this information to increase machinery reliability?

Very simply put, in order to improve the reliability using this method, you have to trust it and use the trends of all the parameters discussed to determine when maintenance is to be performed based on condition, NOT TIME.

In order to do this effectively, the trends for all the components of a train need to be saved in the same folder and be easily accessed to look at all components together. Figure 8.1 discusses this.

Remember, when any deviation in a component condition is noticed, that is the time to be proactive and determine the cause of the change. This can save the plant from extensive maintenance and $$$ from avoiding an unplanned shutdown. When a parameter is in alarm, it is too late!!!

The remainder of this book will provide you with spreadsheets containing all of the parameters we have discussed throughout this book, but putting all of the five major components together for typical types of rotating machinery. The parameters and calculations on these spreadsheets can easily be imported into either your DCS systems or process information systems (PI, ASPEN, or equivalent).

Tables 8.1 through 8.3 contain parameters and calculations for a pump, steam turbine-driven compressor train, and gas turbine to monitor the condition of the five major components.

Please use and implement the information provided in this book, and you will see a huge improvement in your predictive maintenance program, as well as minimal unplanned shutdowns.

Forsthoffer's Component Condition Monitoring
https://doi.org/10.1016/B978-0-12-809599-7.00008-7

Using CCM to reduce PM and optimize reliability

- Set up trends for each of the components on a specific train on a same time basis and have them saved in a folder together for easy access and comparison.
- When a change in one component is observed, immediately review trends of the other four major components to determine what the cause of the issue is.
- Develop an action plan to resolve the issue. If the issue requires maintenance to be performed, determine the rate of deterioration of the component condition, based on the trends and how long it can operate before going into alarm.
- If caught early, maintenance is less likely to be required.
- Delay any maintenance to turnarounds if possible!!!

Fig. 8.1 Using CCM to optimize reliability.

Table 8.1 Pump Component Condition Monitoring

	Pump Component Condition Monitoring			
Performance				
Item #				
Date				
Input				
P_1 (psig)				
P_2 (psig)				
S.G.				
Pump speed (RPM)				
Flow rate (gpm)				
Calculate				
Head (ft)	–	–	–	–
Flow determination[a]				
Control valve position				
CV for valve and trim type[b]				
Fluid S.G.				
Control valve ΔP				
Calculated valve flow	–	–	–	–
Motor amps				
Volts				
Power factor				
Motor eff'y				
Calculated power	–	–	–	–
Flow from pump curve[c]				
T1 (Deg. F)				
T2 (Deg. F)				

Table 8.1 Pump Component Condition Monitoring—cont'd

	Pump Component Condition Monitoring			
CP (specific heat)				
Calculated head	–	–	–	–
Calculated pump Eff'y	–	–	–	–
Flow from pump curve				
Pump maintenance required[d]				
EROE determination				
BEP flow[e]				
EROE min flow	–	–	–	–
EROE max flow	–	–	–	–
Is pump in EROE	–	–	–	–
EROE targets for operations				
Flow				
Amps				
Pump ΔT (measured on inlet and discharge pipes)				

[a]If a flow meter is not available for the pump flow will be determined by one of the alternative methods noted in tab 7 of the Supplementary Manual. These alternative methods are: 1. Portable ultrasonic flowmeter. 2. Control valve position to calculate flow. 3. Motor amps to calculate power. 4. Pump ΔT to calculate pump efficiency.
[b]The CV can be found on the valve manufacturers curve for specific valve type and trim. Note that most of the major valve manufacturers post these curves on their websites.
[c]Flow will be estimated using the original shop test curve. Note that if the pump is not in good condition these estimates will not be very accurate.
[d]Pump maintenance should be considered if the head and flow (operating point) when plotted on the test curve is approx. 10% below test curve flow for calculated head.
[e]Flow at highest efficiency taken from the shop test curve in the Supplementary Manual.

Table 8.2 Steam Turbine-Driven Compressor Train CCM

	Steam Turbine Performance	
	1	2
Item #		
Date		
Given		
P_1 (kPa)		
T1 (Deg. C)		
P_2 (kPa)		
T2 (Deg. C)		
Speed (RPM)		
Flow rate (kg/h)		
Total power at coupling[a] (kW)		

Continued

Table 8.2 Steam Turbine-Driven Compressor Train CCM—cont'd

	Steam Turbine Performance			
	1		2	
Determine				
h1[b] (kJ/kg, from Mollier Diagram)				
h2[b] Isentropic (kJ/kg, from Mollier Diagram)				
Theoretical steam rate (kg/kWh, TSR)	–	–	–	–
KW HP section	–		–	
KW LP section		–		–
h2[b] Actual (kJ/kg, from Mollier Diagram)		–		–
Actual steam rate (kg/kWh, ASR)	–	–	–	–
Steam turbine efficiency	–	–	–	–
% Moisture LP section				
Actions required				

[a]Total power at coupling can be determined via torque meter or from compressor total gas power plus mechanical losses for bearings and seals.
[b]Note that values for h1, h2 isentropic and h2 actual (HP Case only) are obtained from a Mollier Diagram contained in this section or by using steam tables.

Table 8.3 Gas Turbine CCM

	Gas Turbine Component Condition Monitoring						
Equip. #							
Group name							
Compressor performance							
P atmosphere (bara)							
T ambient (C)							
P disch. (bara)							
T disch. (C)							
Speed (RPM)							
K							
$(K-1)/K$	–	–	–	–	–	–	–
$(n-1)/n$	–	–	–	–	–	–	–
Polytropic efficiency	–	–	–	–	–	–	–
What action is required based on % difference from design efficiency?							
Gas turbine performance							
Fuel LHV (kJ/kg)							
Flow (kg/sec)							
Output (kW)							
Site heat rate (kJ/kWh)	–	–	–	–	–	–	–
GT efficiency	–	–	–	–	–	–	–
What action is required based on gas turbine performance info. above?							

Table 8.3 Gas Turbine CCM—cont'd

Gas Turbine Component Condition Monitoring
Bearing condition
1 Journal Brg. Vib X
1 Journal Brg. Vib Y
Major frequency observed
Radial shaft position
1 Brg. pad temp. 1 (C)
1 Brg. pad temp. 2 (C)
1 Brg. drain temp (C)
Thrust Brg. displacement
Direction of thrust
Pad temp. inlet end 1 (C)
Pad temp. inlet end 2 (C)
Pad temp. exhaust end 1 (C)
Pad temp. exhaust end 2 (C)
Thrust Brg. drain temp. (C)
2 Journal Brg. Vib X
2 Journal Brg. Vib Y
Major frequency observed
Radial shaft position
2 Brg. pad temp. 1 (C)
2 Brg. pad temp. 2 (C)
2 Brg. drain temp (C)
Brg. inlet oil pressure (bara)
Brg. inlet oil temperature (C)
Viscosity (cst)
% water in oil
Lube oil flashpoint (C)
What action is required based on gas turbine bearing info. above?
Air filtration system condition
Filter DP (bard)
Action of self cleaning air system (active or inactive)?
Action?
Variable inlet guide vanes
Hyd. supply differential press. (bard)
Guide vane position
Air flow set point
Measured air flow
IGV exhaust temp. reference
Measured exhaust temp. (C)
Action?
Cooling and sealing air system

Continued

Table 8.3 Gas Turbine CCM—cont'd

	Gas Turbine Component Condition Monitoring
Cooling air pressure (bara)	
Other cooling sys. observations	
Air sealing system supply pressure (bara)	
Fuel gas system	
Ambient temp. (C)	– – – – – –
Relative humidity	
Fuel supply pressure (bara)	
Fuel flow rate (kg/h)	
Fuel pressure upstream of shut off valve (bara)	
Fuel shut off valve position	
Fuel pressure upstream of fuel control valve (bara)	
Fuel control valve position	
Fuel supply pressure to combustors (bara)	
Action?	
Combustion monitoring system	
Ambient temp. (C)	– – – – – –
Relative humidity	
Exhaust temp. 1 (C)	
Exhaust temp. 2 (C)	
Exhaust temp. 3 (C)	
Exhaust temp. 4 (C)	
Exhaust temp. 5 (C)	
Exhaust temp. 6 (C)	
Exhaust temp. 7 (C)	
Exhaust temp. 8 (C)	
Exhaust temp. 9 (C)	
Exhaust temp. 10 (C)	
Exhaust temp. 11 (C)	
Exhaust temp. 12 (C)	
Allowable spread (C)	
Top spread 1 (C)	
Top spread 2 (C)	
Top spread 3 (C)	
Exhaust thermocouple alarm on?	
Combustion alarm on?	
High exhaust temp. spread trip?	
Monitor enable activated?	
Action?	
Exhaust temperature control system	
Compressor disch. P (bara)	

Table 8.3 Gas Turbine CCM—cont'd

	Gas Turbine Component Condition Monitoring
Fuel stroke reference	
Ambient temperature (C)	
Relative humidity	
Calculated firing temp. (C)	
Action?	

APPENDIX

Gas Turbine Performance Deterioration and Compressor Washing

Cyrus Meher-Homji*, Andrew F. Bromley†, Jean-Pierre Stalder‡
*Bechtel Corporation, Houston, TX, United States
†Operations Turbotect (USA), Inc., Tomball, TX, United States
‡Turbotect Ltd., Baden, Switzerland

Nomenclature

A/F	air to fuel ratio for the gas turbine
CCPP	combined cycle power plant
CDP	compressor discharge pressure
CDT	compressor discharge temperature
CHP	combined heat and power
Cp	specific heat (J/kgK)
GT	gas turbine
HRSG	heat recovery steam generator
IGV	inlet guide vane
ISO	International Standards Organization
I_f	contaminant level in the fuel (ppmw)
I_{air}	contaminant level in the air (ppmw)
I_w	contaminant level in the injection water (ppmw)
I_{stm}	contaminant level in the injection steam (ppmw)
LHV	lower heating value
NWR	net work ratio, output/W_t
OEM	original equipment manufacturer
PPM	parts per million
PR	pressure ratio
RH	relative humidity
S/F	steam to fuel ratio
T_{as}	static air temperature
Ts	saturation air temperature
TCL	total contaminant level (ppmw)
TIT	turbine inlet temperature
W_c, W_t	compressor and turbine work (kW)
W/F	water to fuel ratio
ΔT **stage**	average total temperature rise/stage (°C)
ηc	compressor efficiency
m	mass flow rate (kg/s)
V	axial compressor volumetric flow rate

Introduction

The use of gas turbines in power generation and other industrial applications has grown significantly in the past two decades. Compressor fouling and its control is, by far, the most important topic in the area of gas turbine performance deterioration and consequently is dealt with in detail. An overview of fouling deterioration may be found in Meher-Homji (1990) and an overall treatment of gas turbine performance deterioration including other sources is presented in Meher-Homji et al. (2001). A detailed treatment of gas turbine degradation is made by Kurz and Brun (2000, 2012), and by Kurz et al. (2008). Flashberg and Haub (1992) have provided a treatment of nonrecoverable deterioration. An overview of fouling deterioration may be found in Meher-Homji and Bromley (2004).

The fouling of axial flow compressors is a serious operating problem and its control is of supreme importance to gas turbine operators especially in the deregulated and highly competitive power market. It is also significant in the mechanical drive market where a loss in gas turbine output directly affects plant throughput. Foulants in the ppm range can cause deposits on blading, resulting in severe performance deterioration. The effect of compressor fouling is a drop in airflow and compressor isentropic efficiency, which results in a "rematching" of the gas turbine and compressor causing a drop in power output and thermal efficiency. In extreme cases, fouling can also result in surge problems, as it tends to move the compressor surge line to the right, i.e., toward the operating line.

Estimates have placed fouling as being responsible for 70%–85% of all gas turbine performance losses accumulated during operation. Output losses between 2% (under favorable conditions) and 15%–20% (under adverse conditions) have been experienced.

A treatment of fouling susceptibility and sensitivity was made by Meher-Homji et al. (2009) and key findings are incorporated here. It is useful to recognize that there are two distinct issues to be considered:

- The susceptibility of different gas turbine designs to fouling, i.e., the axial compressor's propensity to foul.
- The **sensitivity** of different gas turbine designs to the impact of fouling, i.e., the effect of fouling on their performance.

In the past, these two issues have often been wrongly analyzed or confused. Fouling susceptibility and the effect of fouling are quite different. Results of

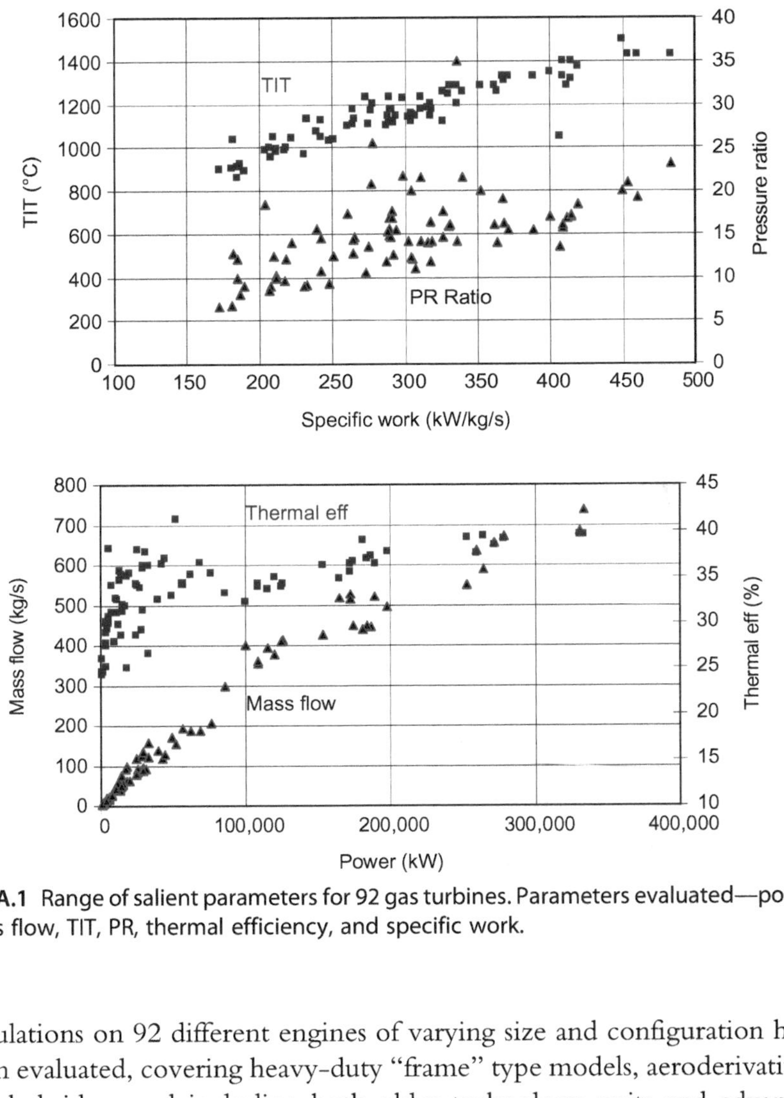

Fig. A.1 Range of salient parameters for 92 gas turbines. Parameters evaluated—power, mass flow, TIT, PR, thermal efficiency, and specific work.

simulations on 92 different engines of varying size and configuration have been evaluated, covering heavy-duty "frame" type models, aeroderivatives, and hybrids – and including both older technology units and advanced models with a wide span of characteristics. The range of salient parameters used for the analysis is shown in Fig. A.1.

Gas Turbine Airflow Ingestion and Fouling

Gas turbines ingest extremely large quantities of air, with larger gas turbines having airflow rates as high as 680 kg/s (1500 lbs/s). A scatter plot

Table A.1 Example of Airflow Ingestion of Different Gas Turbines and Ratio of Compressor Work to Total Turbine Work

Gas Turbine	ISO Output (kW)	ISO Airflow (lb/s)	Air Ingested Per Year (Miles Above a Football Field)	Foulant Ingested Per Year at 10 ppm Ambient (lb)	Compr. Work W_c (kW)	Total Turbine Work W_t (kW)	W_c/W_t	Net Work Ratio (NWR)
Typhoon	5106	39	53	12,299	8104	13,536	0.59	0.41
Centaur 50	4481	42	57	13,182	6263	11,093	0.56	0.44
Mars 100	10,436	92	126	29,013	17,938	28,981	0.61	0.39
Frame 5371 PA	26,056	269	367	84,832	38,960	66,311	0.59	0.41
RB-211	26,678	199	272	62,757	41,222	69,258	0.60	0.40
Frame 6581B	41,487	321	438	101,231	50,706	94,229	0.54	0.46
Trent 50	50,990	340	464	107,222	84,280	136,478	0.61	0.39
GT 8C2	56,295	429	585	135,289	77,627	136,149	0.57	0.43
Frame 6101FA	70,116	450	614	141,912	74,237	147,127	0.50	0.50
Frame 7121EA	85,206	655	894	206,561	105,264	192,481	0.55	0.45
GT 13E2	160,748	1153	1573	363,610	190,196	354,064	0.54	0.46
Frame 7241FA	172,757	971	1325	306,215	164,387	340,671	0.48	0.52
Frame 9351FA	255,400	1429	1950	450,649	229,707	490,325	0.47	0.53

of the airflow rate versus power for 92 gas turbines is presented in Fig. A.1. A very important parameter in evaluating gas turbines is "specific work"—defined as the power per unit of airflow rate (kW/kg/s). The linkage between airflow rate and compressor fouling is indicated in Table A.1. The data were derived by GTPRO[1] simulations running at an ambient temperature of 15°C (59°F) and with typical inlet and outlet losses. This table indicates the ingested amount of foulant assuming an ambient loading of 10 ppm for a variety of gas turbines.

Axial compressor work [W_c] and total turbine work [W_t] are also shown in Table A.1, and it can be seen that the compressor section consumes a large portion of the total turbine work. To quantify this important fact, the last column of Table A.1 provides the W_c/W_t ratio and it can be seen that

[1] By Thermoflow Inc.

~50%–60% of the total work produced in the turbine is consumed by its axial compressor. Consequently, maintaining high compressor efficiency is important for the plant's revenue stream.

To help visualize the huge airflow, the volume of air consumed per year is presented in terms of miles above a traditional football field 110 × 49 m (360 × 160 ft). As an example, the Frame 9351FA gas turbine, (ISO airflow of 648 kg/s), would ingest, in a year of operation, a column of air over a football field 3120 kM (1950 miles) high. At a 10 ppm foulant loading rate, 204,414 kg (450,649 lb) of foulant would be ingested.

Solids or condensing particles in the air and in the combustion gases can precipitate on the rotating and stationary blades causing changes in aerodynamic profile, reducing the compressor mass flow rate and affecting the flow coefficient and efficiency, thus reducing the unit's overall performance. Further, contaminated air can cause a host of problems that include erosion, fouling, corrosion and, in some cases, plugging of hot section cooling passages. There is also a close correlation between mechanical reliability and fouling deterioration, and an example is the damaging effects of fouling on blading integrity as discussed in the following sections. This is another important reason to keep the compressor clean. Some typical photos of fouled compressors are shown in Figs. A.2–A.4.

Fig. A.2 Examples of fouled compressor blades: (A) heavily fouled air inlet bell-mouth and blading on a 35 MW gas turbine, (B) typical oily deposits on blades, and (C) compressor blades fouled with a mixture of salts and oil.

Fig. A.3 Oily deposits on axial compressor blades from no. 1 bearing oil leakage on a large heavy duty gas turbine.

Fig. A.4 Salt deposits on compressor blades. *(Courtesy, Turbotect Ltd.)*

Economic Impact of Fouling

The cost of performance deterioration varies from site to site, depending on the rate of fouling and the effectiveness of compressor washing programs that are applied. Actual costs can be significant, and are often underestimated by operating companies.

Basically, fouling cost calculations must address the value of the resulting shortfall in energy output and the cost of fuel that is consumed over the operating period. The manner by which the output shortfall is measured may need to be adjusted according to the specific application. For example, in power generation it is the lost revenue from electricity sales, whereas in mechanical-drive gas transmission applications, the calculation will involve the quantity of gas moved through the pipeline and the sales price of the gas. Cost models for combined cycle plants must also address the impact of gas turbine output deterioration on the performance of the steam cycle, and these calculations can become more complex as a result of single- and multi-shaft configurations, etc.

Gas turbine operating mode also affects the cost calculation. For example, when operating at or near to ***base load***, power output degradation can usually not be controlled by the consumption of additional fuel—due to turbine firing temperature control limitations. In this case, and for a power generation application, the cost of compressor fouling is driven by the loss of revenue from electricity sales, and is partially offset by the cost of fuel saved due to the reduced power output.

However, under ***part load*** operating conditions (and within certain limits), power output degradation due to compressor fouling can be controlled by consuming additional fuel. In this case, and assuming the power output is fully recovered, there will be no loss of revenue from electricity sales and the cost of degradation becomes the cost of the additional fuel.

Fig. A.5 shows the results of a realistic fouling cost model applied to three large, heavy-duty gas turbines in a power generation application. It assumes 8000 h operation at base load, representative prices for fuel purchase and power sales, and typical "modest" fouling conditions resulting in a 5% reduction in power output at the end of the period. Note that the area above the degradation curve computes the annual power output shortfall (in MWh), which in this example calculates to a loss of 3.83%. After allowing for fuel cost savings due to reduced power output, the estimated net cost degradation for these three engines ranges from about 1.3 to 4.6 million US dollars per year.

Power losses attributable to compressor fouling can be restored through regular compressor cleaning and judicious plant maintenance programs. Overall plant profitability can be significantly improved for a relatively small cost. The amount of improvement at a given site depends on the type of cleaning program adopted and the thoroughness of its implementation. All elements of the program are important, including the design of

This example assumes 8000 h operation at 100% base load
Power output loss at end of period = 5.0%
NG fuel purchase price = 3.0 US$/MMBtu
Power sales price = 0.07 US$/kWh

Based on typical asymptotic degradation curve, representing a medium fouling rate →

Model also assumes a typical heat rate increase of about 0.4% per 1% power output loss

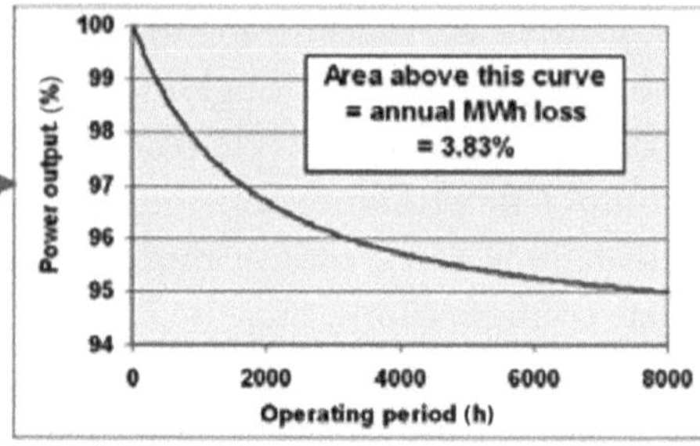

GT type		Fr7EA	Fr7FA	V94.3A
Power output, clean	MW	85.0	172.0	285.0
Power o/p at end of period	MW	80.8	163.4	270.8
Net cost of degradation	$MM	1.268	2.685	4.556

Fig. A.5 Annual cost of compressor fouling, estimated for three heavy-duty gas turbines (Bromley, 2012).

compressor cleaning systems such as wash skids and injection nozzles, the choice and use of detergents, the frequency of cleaning, and the actual washing procedure used.

Aerodynamic Considerations

An axial compressor is a machine where the aerodynamic performance of each stage depends on that of the earlier stages. Thus, when fouling occurs in the inlet guide vanes and the first few stages, there may be a dramatic drop in compressor performance. This can often occur when oil and industrial smog or pollen are present and form adherent deposits. The forward compressor stages are usually fouled the worst. If the rear stages foul, this seems to have a smaller impact on performance; but due to higher temperatures, deposits can become baked and difficult to remove. This baking effect is more severe on the high pressure ratio compressors of aeroderivative machines ranging from 18:1 to 35:1 pressure ratio, as opposed to the typical 10:1 or 14:1 pressure ratios found on the heavy duty industrial gas turbines.

Fig. A.6 shows the changes in compressor efficiency and heat rate occurring in a large gas turbine over time.

A typical characteristic curve for an axial flow compressor stage is shown in Fig. A.7. Under design operating conditions, most stages would operate at design flow coefficient and at a high isentropic efficiency. When the flow coefficient is to the right of the characteristic curve, the stage is lightly loaded and the extreme right point is known as the choke point. To the left of the characteristic curve is a region where aerodynamic stall occurs (surge region).

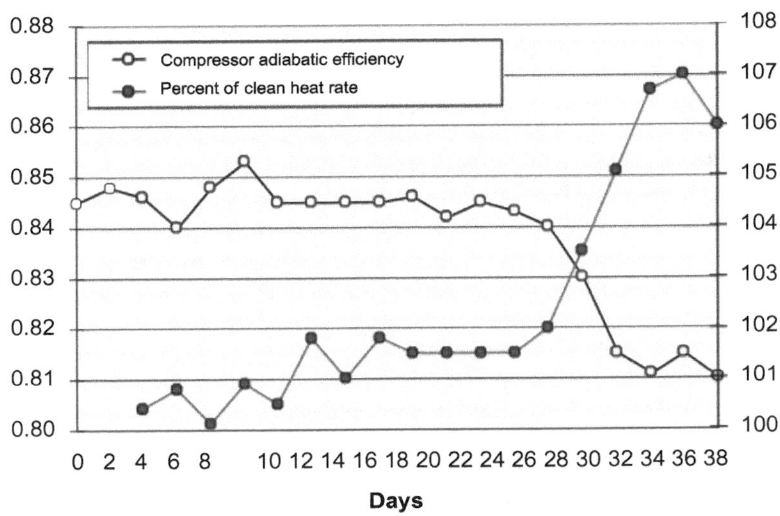

Fig. A.6 Change in compressor efficiency and heat rate for a heavy duty gas turbine.

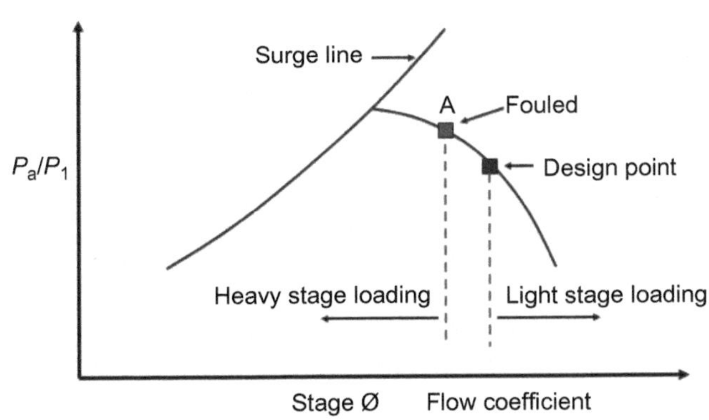

Fig. A.7 Compressor stage loading characteristics.

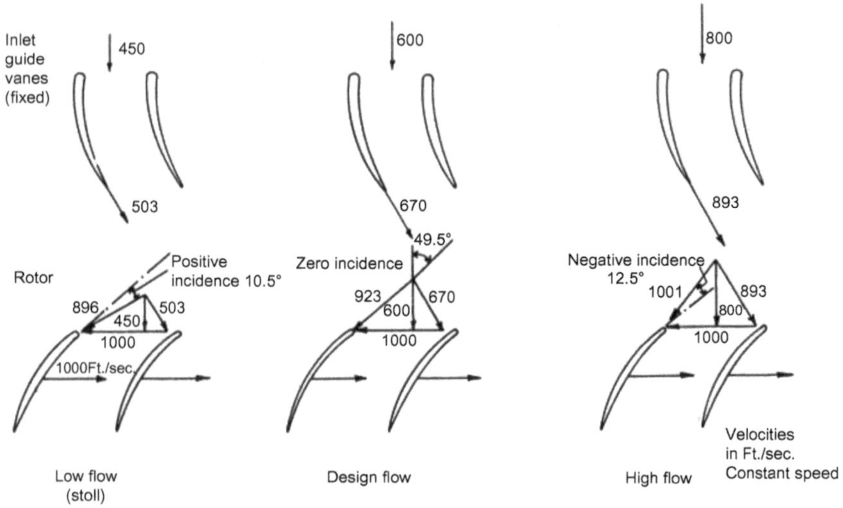

Fig. A.8 Velocity triangles under fouled conditions showing how change in mass flow causes a change in incidence (Dundas, 1982).

As fouling drops the mass flow (flow coefficient) in the first stage, this affects the performance of the latter stages as follows: The operating point on the first-stage characteristic moves toward the left, thus increasing the pressure ratio. This causes a higher density at the inlet to the second stage. Thus there will be a further reduction in second-stage flow coefficient. This effect progresses through successive stages until aerodynamic stall occurs in a later stage and triggers a surge. Basic velocity triangles indicating how a drop in mass flow causes excessive incidence angles and subsequent aerodynamic stall are shown in Fig. A.8 (Dundas, 1982).

Simulation Results of Compressor Deterioration

Simple Cycle Simulation

It is instructive to investigate the sensitivity of compressor fouling deterioration on simple cycle gas turbine performance. To this end, simulation runs using GTPRO software have been made for a 40 MW-Class heavy duty gas turbine in simple cycle configuration. Runs were made with natural gas fuel with an LHV of 50,046 kJ/kg (21,518 Btu/lb) and with typical inlet and outlet pressure drops 100 and 125 mm WG (4 and 5 in. WG), respectively. The machine has an ISO pressure ratio of 11.8:1 and a mass flow rate of 138 kg/s (304 lbs/s) and a firing temperature of 1104°C (2020°F).

The simulation was run at an ambient temperature of 15°C (59°F), and imposing deterioration steps in the following sequence:

- Step 1: New and clean, mass flow drop = 0%, comp. efficiency drop = 0%
- Step 2: Mass flow drop = 1%, comp. efficiency drop = 0.833%
- Step 3: Mass flow drop = 2%, comp. efficiency drop = 1.67%
- Step 4: Mass flow drop = 3%, comp. efficiency drop = 2.5%
- Step 5: Mass flow drop = 4%, comp. efficiency drop = 3.33%
- Step 6: Mass flow drop = 5%, comp. efficiency drop = 4.167%
- Step 7: Mass flow drop = 6%, comp. efficiency drop = 5%

Output and heat rate variation with the deterioration steps are shown in Fig. A.9. The output at the end of the 7th deterioration step has dropped 5.5 MW while the heat rate has increased by 850 Btu/kWh. The change in mass flow rate, compressor discharge pressure, and compressor discharge temperature corresponding to the simulated deterioration steps is shown in Fig. A.10.

The drop in efficiency causes the discharge temperature to increase by ~10.6°C (19°F) and the compressor discharge pressure to drop by about 0.7 bar (10 psia). The variation in axial compressor work, turbine section work, and the output after losses is shown in Fig. A.11.

Whereas the axial compressor work is seen to drop slightly due to the reduction in mass flow (middle line of the figure), there is a steep drop in

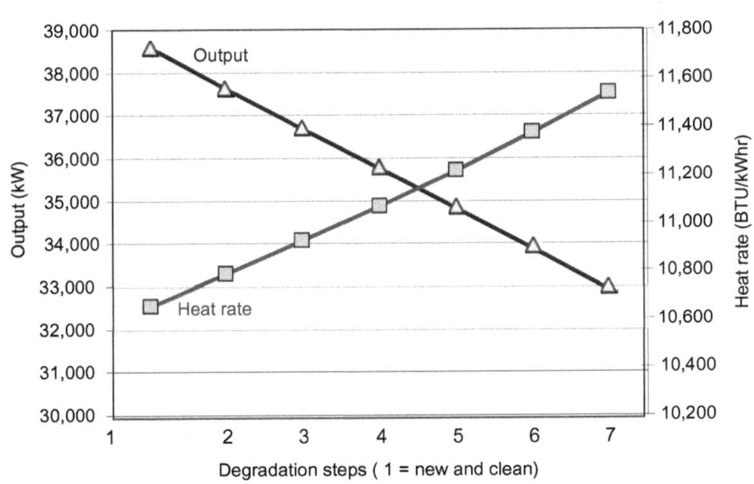

Fig. A.9 Output and heat rate change with compressor degradation steps imposed on a 40 MW gas turbine.

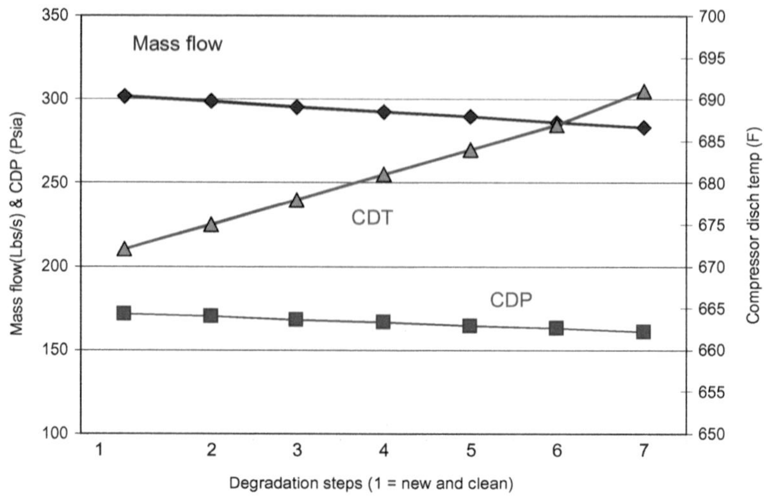

Fig. A.10 Change in mass flow rate, compressor discharge temperature, and pressure with compressor degradation steps imposed on a 40 MW gas turbine.

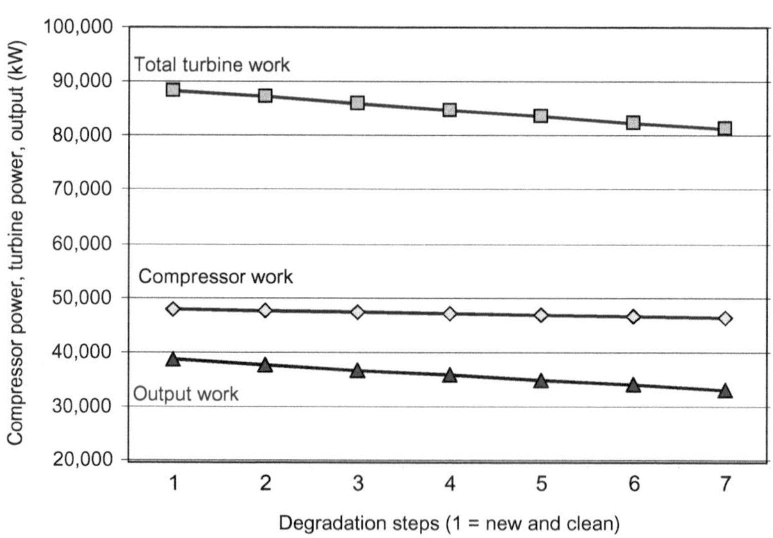

Fig. A.11 Change in compressor work, turbine work, and overall output with compressor degradation steps imposed on a 40 MW gas turbine.

the turbine work (upper line in the figure) resulting in a drop in overall gas turbine output of 5.5 MW. The steep drop in turbine section work is due to the reduced mass flow and the smaller expansion ratio available as a result of the loss in compressor discharge pressure.

Even with good air filtration, salt can collect in the compressor section, and will continue to accumulate (together with other foulants) until an equilibrium condition is reached. At this point, large particles will start to break away and will be ingested into the combustion section in relatively high concentrations. This ingestion has to be prevented by the removal of salt from the compressor prior to saturation. The rate at which saturation occurs is highly dependent on filter performance.

The effect of hot section fouling is that the nozzle throat area is reduced. As this controls the compressor–turbine match, it causes a movement away from the design match point and results in a corresponding loss of performance. Deposits will also form on the rotating blades causing a further loss in performance.

Also, as the turbine section of a gas turbine fouls, there will be a drop in the turbine flow coefficient and the compression ratio of the compressor will increase as the turbine "swallowing" capacity is reduced. Note that in some OEM control systems, the compression ratio and exhaust gas temperature are used to determine the turbine inlet temperature; Zaba (1980). This algorithm is based on an assumption of constant turbine efficiency, and if this efficiency decreases due to turbine fouling, then the control system will indicate a higher turbine inlet temperature than is really present. Therefore the output of the turbine is further reduced.

Combined Cycle Simulation

In order to examine the effect of compressor fouling on combined cycle plant, simulations were conducted on a combined cycle power plant (CCPP) based on a 7241FA gas turbine operating with a three-pressure level HRSG and a reheat condensing steam turbine. The gas turbine is ISO rated at 174 MW and operates at a pressure ratio of 15.5:1 and a firing temperature of 1327°C (2420°F) The ISO mass flow rate is 448 Kg/s (988 lbs./s). Typical inlet and outlet losses for a CCPP were considered to be 100 and 254 mm WG.

The same seven deterioration steps were considered as for the simple cycle model above, and the effect of compressor fouling deterioration on CCPP output power and heat rate are indicated in Figs. A.12 and A.13. As can be seen in Fig. A.12, the drop in gas turbine output and the slight reduction in steam turbine output result in a net power drop of ~23.7 MW. As the gas turbine flow drops, the exhaust gas temperature increases, thus resulting in a relatively moderate drop in steam turbine performance of 2.3 MW. The

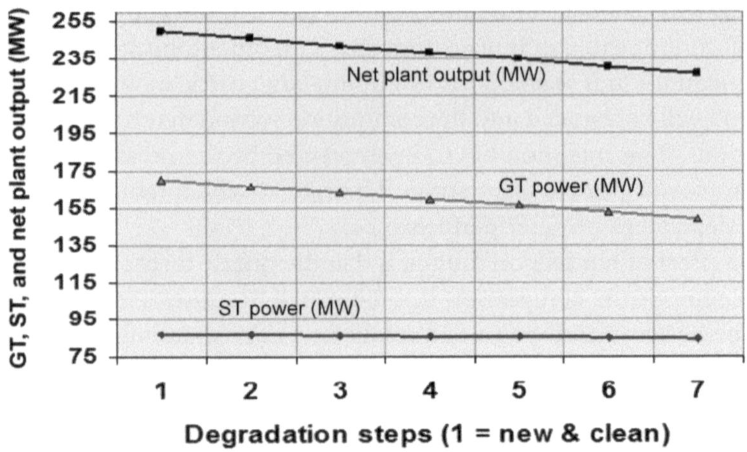

Fig. A.12 Change in net power, GT output and ST output in a combined cycle with compressor degradation steps.

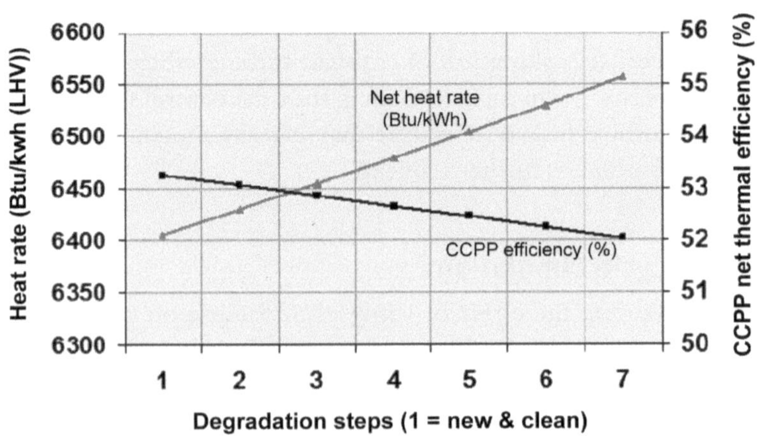

Fig. A.13 Change in net cycle efficiency and heat rate with compressor degradation steps.

heat rate increase and overall CCPP efficiency drop are depicted in Fig. A.13, and it can be seen that the drop in the overall CCPP efficiency is almost 1.22%.

To examine the fouling sensitivity of a variety of gas turbines, a number of simulations were run using GTPRO software. The analysis included 92 gas turbines, including heavy duty and aeroderivative engines covering a range of operating parameters of pressure ratio, turbine inlet temperatures, and specific work shown in the scatter plots of Fig. A.1.

All runs were done at 15°C and 40°C at sea level, and at a relative humidity of 60%. Fuel used was CH_4 with LHV 50,046 kJ/kg. All runs were made at base load using an inferred TIT control model. The inlet and exit losses were 10 and 12.45 mbar, respectively. The approach followed is described below:

- Firstly, runs were made on all gas turbines at 15°C under new and clean conditions. Runs were then made with an imposed level of compressor deterioration. The deterioration was modeled by imposing a reduction in mass flow of 5% coupled with a compressor section efficiency reduction of 2.5% points. These values are considered normal and can be expected to provide a relative comparison of the 92 gas turbines considered.
- Runs as defined in point (1) above were then made at an ambient temperature of 40°C to examine the impact of ambient temperature.
- All gas turbine salient operating parameters were logged into a spreadsheet for analysis.
- Using the spreadsheet, computations were then made of deterioration levels in terms of power and heat rate, and other parameters such as the temperature differential per stage and fouling factors.

Key Gas Turbine Design Parameters

Key design operating parameters used to examine what correlations existed between the loss of power due to fouling included pressure ratio, specific work, and GT net work ratio (NWR). Of these, it appears that the best correlation obtained was with the GT net work ratio (Meher-Homji, et al. 2009). The NWR is defined as the useful output of the gas turbine divided by the total turbine work.

$$\text{NWR(Net work ratio)} = \frac{\text{Output in kW}}{W_t}$$

Another way of expressing this ratio is

$$\text{NWR} = \frac{W_t - W_c}{W_t} = \frac{1 - W_c}{W_t}$$

NWR requires knowledge of the turbine section total work and the work consumed in the compressor, and these are factors that are not easily available to most users.

Power, Heat Rate, and Fuel Consumption Changes

We now examine fouling behavior (i.e., sensitivity to a certain imposed fouling) for the 92 gas turbines. The reduction in power vs the base output of the gas turbine (a rough analog of its size) is shown in Fig. A.14. The traditional anecdotal observation that smaller machines are more sensitive to fouling is supported by these results.

The percent power reduction can also be seen to be inversely correlated to the net work ratio as shown in Fig. A.15, with engines having low net work ratios tending to be more sensitive to a given amount of imposed fouling.

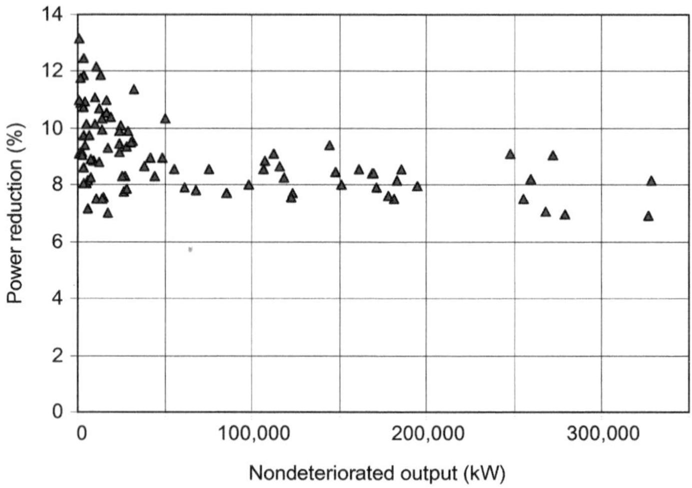

Fig. A.14 Percent power reduction (due to imposed fouling) vs gas turbine nondeteriorated output in kW (~ size).

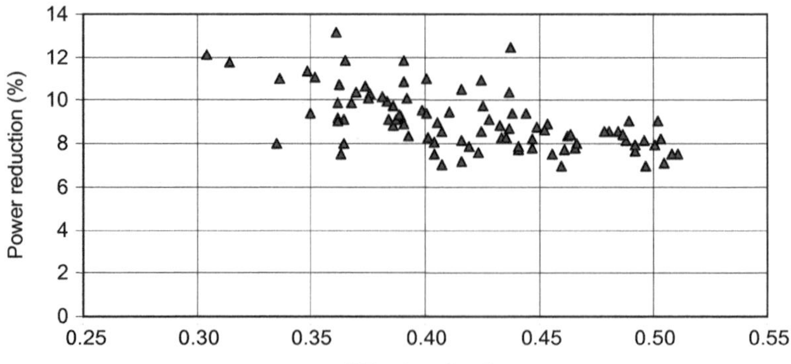

Fig. A.15 Power reduction due to fouling vs net work ratio.

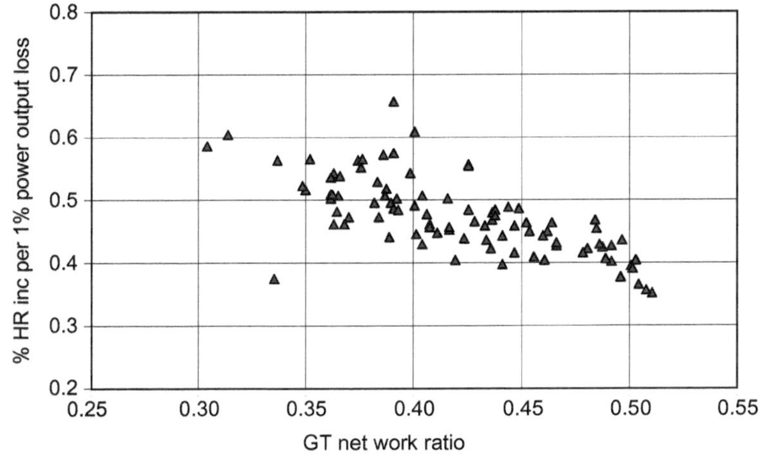

Fig. A.16 Percent heat rate increase per 1% power output reduction (due to fouling) vs net work ratio.

The behavior of heat rate with fouling deterioration is shown in Fig. A.16. This plot provides a rough estimate of the relationship of heat rate deterioration with a certain amount of fouling for different net work ratio engines. Note that E and F class engines operate at a NWR of between 0.4 and 0.5, and show a ratio change in heat rate of about 0.4%–0.5% per 1% deterioration in power output. The heat rate increase due to fouling deterioration for a 1% power output reduction is shown in Fig. A.17 as a function of the gas turbine output.

If a gas turbine is operating in a fouled condition at base load, then the fuel consumption under fouled conditions will actually be reduced as is shown in Fig. A.18. With some gas turbines (and depending on the control mode), it might be possible to increase the fuel flow to mitigate the effect of fouling.

Ambient Temperature Effects

The general sensitivity of power drop with ambient temperature is depicted in Fig. A.19, in which the power drop per degree centigrade (within the range 15–40°C) is plotted against the design gas turbine net work ratio. This curve is without any imposed fouling degradation. As expected, as the compressor work increases as a fraction of the turbine total work, the net work ratio decreases. Thus, low net work ratio machines tend to exhibit higher sensitivity to ambient temperature swings. This is typical of aeroderivative

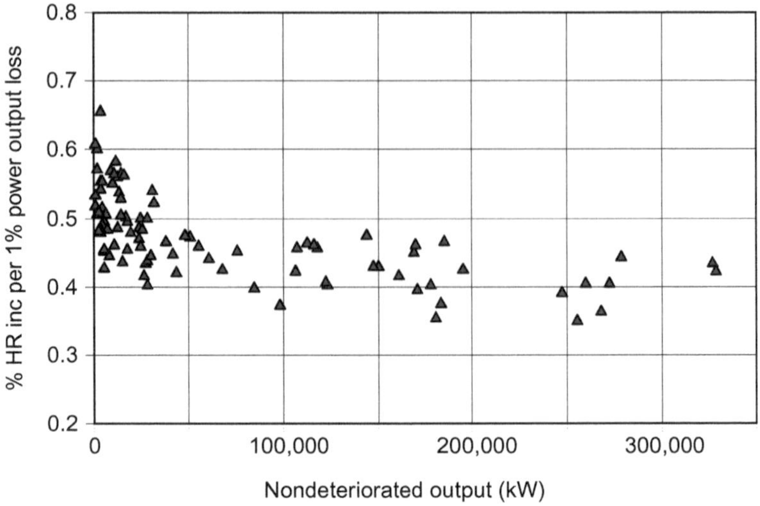

Fig. A.17 Percent heat rate increase per 1% power output drop with 5% reduction in air mass flow (due to fouling) vs gas turbine nondeteriorated output in kW (~size).

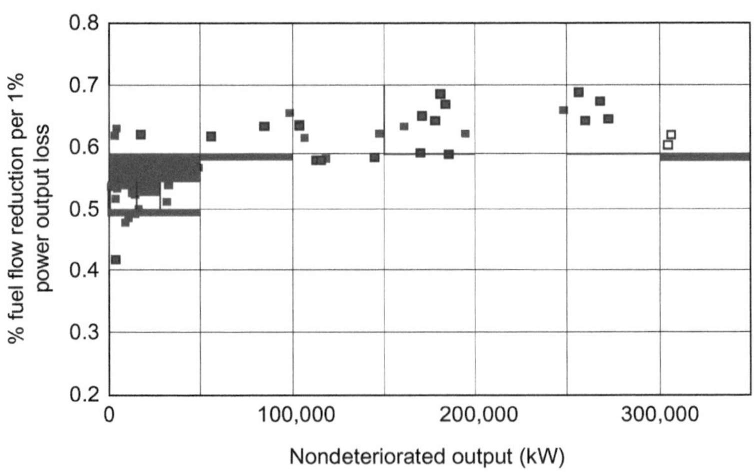

Fig. A.18 Percent reduction in fuel flow rate per 1% output reduction with fouled operation (5% drop in mass flow).

engines, where the compressor work tends to be a higher percentage of the total turbine work.

The power reduction with fouling deterioration vs gas turbine net work ratio for two ambient temperatures of 15 and 40°C is shown in Fig. A.20. It can be seen that the effect of fouling is a function of the net work ratio and

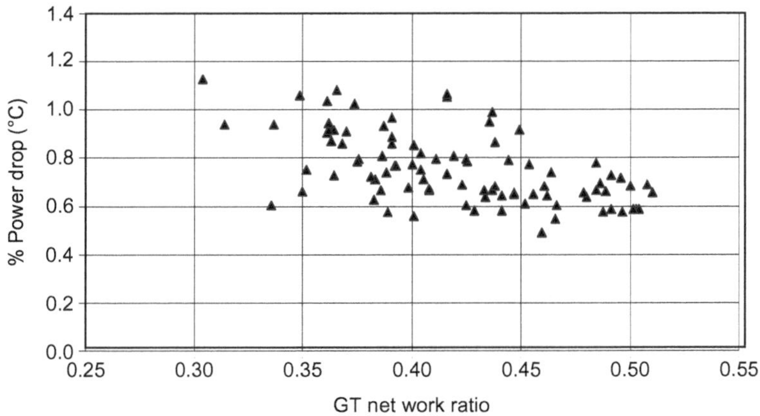

Fig. A.19 Power drop per °C for 92 gas turbines, as a function of GT net work ratio.

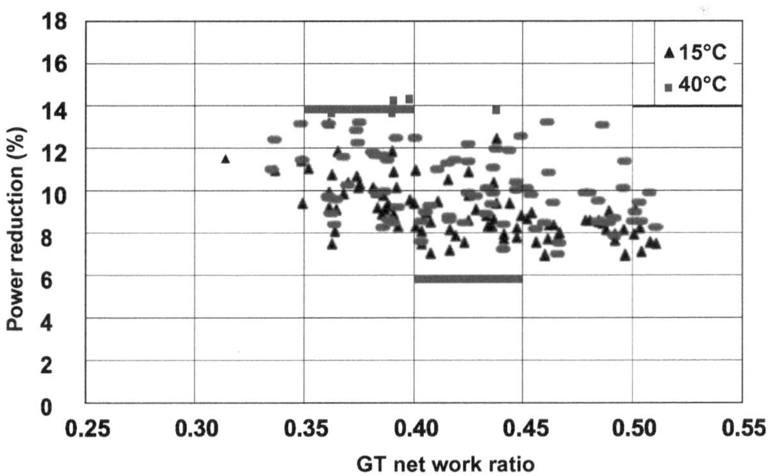

Fig. A.20 Power reduction with compressor deterioration for 92 gas turbines, as a function of GT net work ratio for two different ambient temperatures (15°C and 40°C).

that it is also more severe at high temperatures. A similar plot showing the heat rate change is provided in Fig. A.21.

Key Parameter Changes With Compressor Fouling

A simulation was conducted on a large advanced gas turbine to demonstrate key parameter changes resulting from progressive compressor fouling. The imposed degradation steps and the resulting percentage change of key

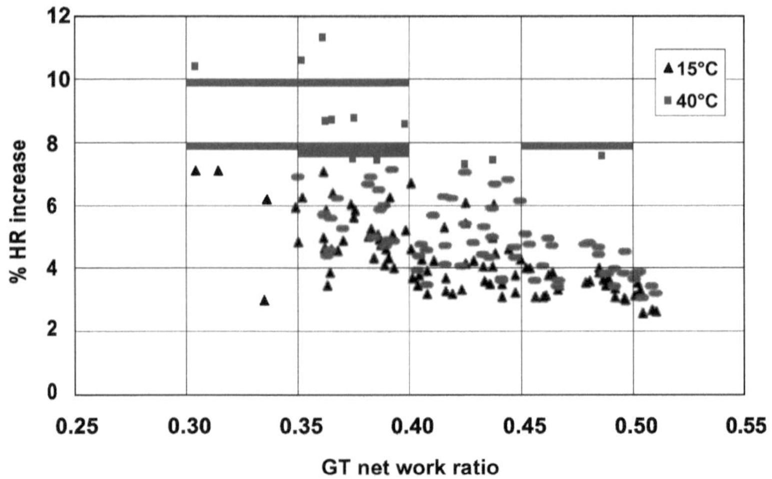

Fig. A.21 Heat rate increase with compressor deterioration for 92 gas turbines, as a function of GT net work ratio for two different ambient temperatures (15°C and 40°C).

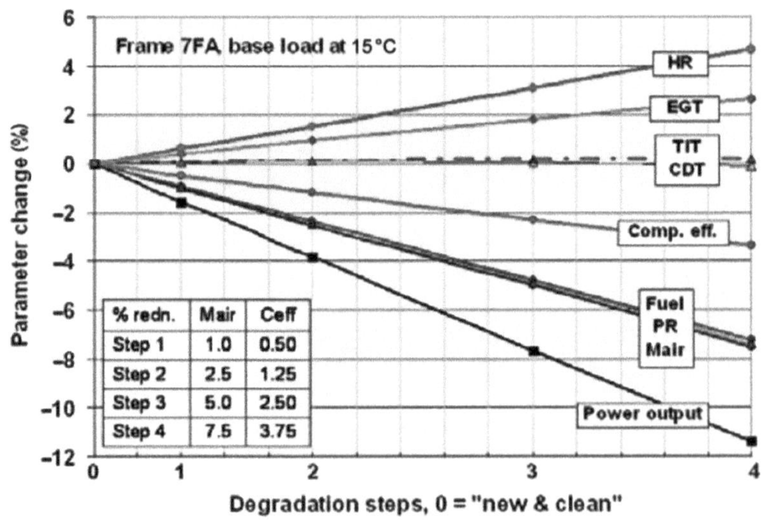

Fig. A.22 Key parameter changes with simulated compressor degradation steps (Bromley, 2012).

parameters are summarized in Fig. A.22. The simulation model assumes base load operation on natural gas at 15°C ambient temperature, and degradation step 0 represents "new and clean" conditions.

It is important to note that the absolute fuel flow rate actually decreases with increasing levels of fouling, while the heat rate increases. In other words, less fuel is consumed due to the declining power output, but it is consumed less efficiently. As discussed in an earlier section on the economic impact of compressor fouling, the relationship between power output, heat rate, and fuel consumption should not be overlooked when calculating the net cost of degradation.

Underlying Causes of Fouling

Experience has shown that axial compressors will foul in most operating environments—be they industrial, rural, or marine. There are a wide range of industrial pollutants and a range of environmental conditions (fog, rain, humidity) that play a part in the fouling process.

Compressor fouling is typically caused by

- Airborne salt
- Industrial pollution—hydrocarbons, fly ash, smog, exhaust emissions from traffic, etc. This causes a grimy coating on the early compressor stages and can get "baked on" in the latter stages because of the high compressor discharge temperatures (this is especially true of high pressure ratio compressors)
- Ingestion of gas turbine exhaust or lube oil tank vapors
- Mineral deposits such as limestone, coal dust, and cement dust
- Airborne materials—soil, dust, sand, chemical fertilizers, insecticides, and plant matter
- Insects—This can be a serious problem in tropical environments.
- Internal gas turbine oil leaks—Leakage from the front bearing of the axial compressor is a common cause. Oil leaks combined with dirt ingestion cause heavy fouling problems.
- Impure water from evaporative coolers (carryover)
- Spray paint that is ingested
- Vapor plumes from adjacent cooling towers

Often the inlet struts and IGVs become severely fouled. Hand cleaning the IGVs and first stage will restore a considerable amount of performance.

Ambient air can be contaminated by solids, liquids, and gases. Air loadings can be defined in mg/m^3, grains/1000 ft^3 or ppm (mass of contaminant per unit mass of air). In general, particles up to 10 μm cause fouling, but not erosion. Particles above 10–20 μm cause blading erosion. Some typical air loadings are as follows:

- Rural 0.01–0.1 ppm by weight
- Coastal 0.01–0.1 ppm by weight

- Industrial 0.1–10 ppm by weight
- Desert 0.1–700 ppm by weight

Felix and Strittmatter (1979) have detailed the type of analysis that should be done at a gas turbine plant site. In most industrial areas, the air quality can create quite acidic conditions in the axial compressor.

The importance of climatic conditions, rain showers, relative humidity, etc. cannot be overemphasized. Several operators have reported dramatic drops in gas turbine output coincident with rain showers. Often air filters will exhibit a sudden growth in differential pressure as the filters get saturated with water due to high humidity. Under certain conditions, the filter may suddenly unload into the airflow causing rapid compressor fouling.

The susceptibility of different types of gas turbines to compressor fouling (as a function of their design parameters) has been studied by Seddigh and Saravanumuttoo (1990). Aker and Saravanumuttoo (1988) have also provided results pertaining to fouling based on stage stacking techniques. More recent findings by Tarabrin and Stalder are presented ahead.

Importance of Recognizing Site Specifics for Fouling Control

It is very important to emphasize the importance of site specifics that will impact the extent and severity of fouling and the wash control method. While this may seem axiomatic, there have been several dogmatic positions taken in the gas turbine community regarding compressor washing, based on findings at one site (or with one type of equipment) that are then generalized to global applications. This causes a lot of controversy with respect to topics such as the choice of cleaning fluids, frequency of washes or the efficacy of on-line water washing. Several OEM water wash systems in the past were inadequately designed and have resulted in poor performance. This sort of situation often leads to polarized positions within even the same operating companies with respect to choice of cleaners, or use of on-line washing. Part of the objective of this paper is to provide a comprehensive body of information, pointing out that each application is unique, and an optimal compressor wash strategy has to be worked out.

Marine and Offshore Environment

The offshore environment is particularly challenging. Some key issues are

- Airborne salt can exist in three basic forms: aerosol, spray, and crystal. Aerosols can range in size from 2 to 20 μm ($1\,\mu m = 10^{-6}$ m) and are generated by bubbles shattering on the sea surface. Sea spray generates large droplets sized 150–200 μm and these tend to drop out due to gravity. Sea salt crystals absorb moisture under appropriate relative humidity conditions, and sizes peak in the range of 2 μm. The relative humidity offshore was found to be almost always high enough to ensure that salt was in its wet form. Studies by Tatge et al. (1980) concluded that salt would stay as supersaturated droplets unless the relative humidity dropped below 45%.

- The environment on offshore platforms is not "dust free" and can include flare carbon, drilling mud, cement, and other dusts. These can pose a problem with poorly positioned flare stacks and with sudden changes in wind direction. Grit blasting has also been a serious problem.

There is considerable variance in data regarding the amount of salt present in an offshore environment. Airborne salt concentrations provided by different researchers are shown in Fig. A.23 (Cleaver, 1988). The NGTE "30 knot standard" is often used to define the environment of offshore platforms.

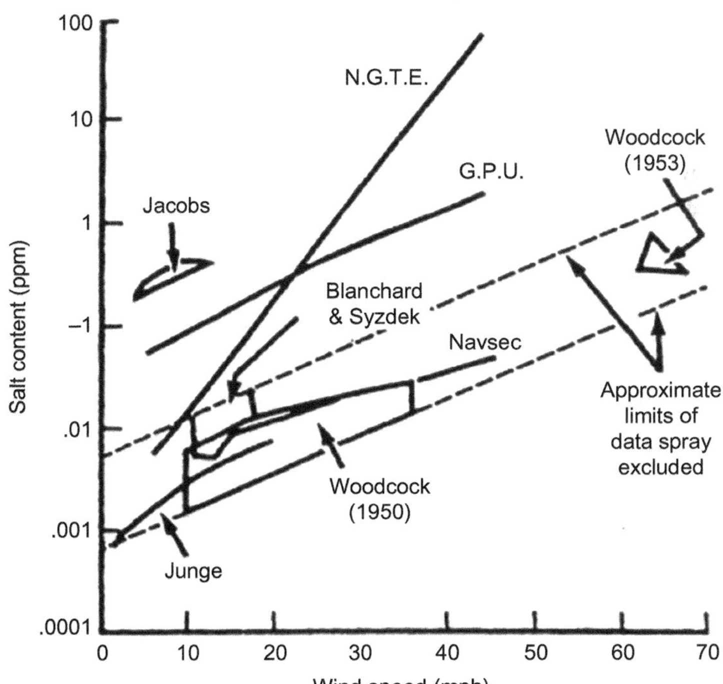

Fig. A.23 Airborne salt concentrations (Cleaver, 1988).

Fouling Deterioration Rate in Axial Flow Compressors

The types of foulants entering the compressor vary widely from site to site, and influences selection of wash detergent as discussed in a later section of this paper. Deposits of oil and grease are commonly found in industrial locations as a result of local emissions from refineries and petrochemical plants. These types of deposits act as "glue" and entrap other materials entering the compressor. Coastal locations usually involve the ingestion of sea salt, desert regions attract dry sand and dust particles, and a variety of fertilizer chemicals may be ingested in agricultural areas.

Compressor foulants are often classified as being "oil-soluble," "water-soluble," or "water-wettable," but experience has shown that they typically are a combination of these types. For example, although sea salt is essentially water-soluble, its retention within the compressor may be significantly influenced by trace quantities of oil and grease. In this case, the use of water alone for washing may not be sufficient, and a chemical detergent would be required for effective compressor cleaning. The susceptibility of gas turbine axial flow compressor to fouling is controlled by the following major factors (Stalder 1998).

- Location of the plant and the environment
- Atmospheric parameters
- Gas turbine design parameters
- Plant design and layout
- Plant maintenance practices

A discussion of fouling susceptibility and sensitivity may be found in Meher-Homji et al. (2009). In practice while design parameters may impact fouling susceptibility, the degree and efficiency of air filtration are of critical importance.

Fouling Degradation Rate

Experience has shown that faster fouling rates occur during initial operation and tend to reduce over time. One particular study by Tarabrin et al. (1996) reported a classical exponential degradation trend, where the fouling rate stabilized after 1000–2000 h.

In practice, all sites have unique fouling conditions that result in different degradation trends ranging from almost linear to exponential, as illustrated in Fig. A.24.

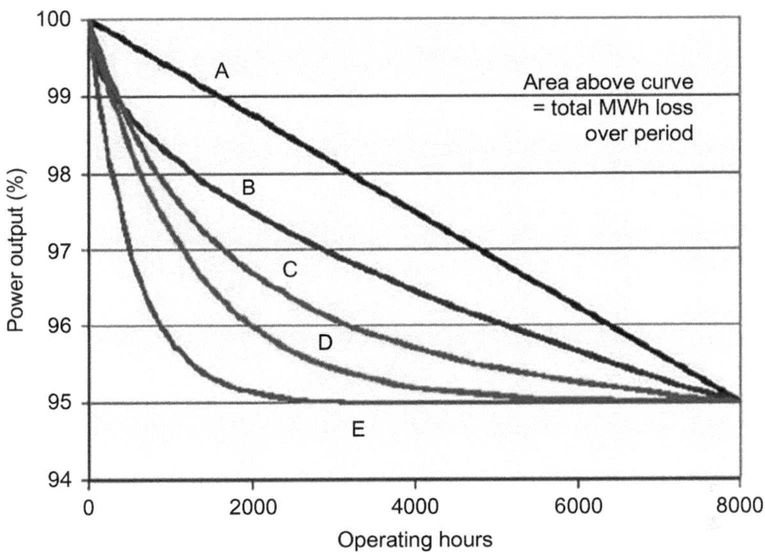

Fig. A.24 Illustration of different degradation trends (Bromley, 2012).

In the example illustrated in Fig. A.24, power output drops by 5% over 8000 h, but the actual power generation shortfall (MWh) over the operating period is significantly different for each degradation curve, i.e., it is a function of the area above the curve. For example, the linear curve (*A*) gives the smallest annual loss of 2.5% MWh, whereas the exponential curve (*E*) results in a much higher annual loss of about 4.65% MWh.

Location of Fouling in an Axial Compressor

Experimental studies conducted by Tarabrin et al. (1998) have indicated that the first 5–6 stages of the compressor tend to foul, and the degree of fouling tends to decrease from the front to the back end of the compressor. Deposits on the blades measured from a 16-stage frame 5322 gas turbine are quantified in Fig. A.25 for the rotor blades and in Fig. A.26 for the stator blades.

These figures provide the weight distribution on the convex and concave sides of the rotor and stator blades up to stage 6. An insignificant amount of deposits were noted on the seventh stage onward.

Relationship Between Airflow Loss Due to Fouling, Pressure Ratio and Gas Turbine Thermal Efficiency

For single-shaft machines, the percent change in mass flow rate due to compressor fouling and pressure ratio are approximately equal. The relationship

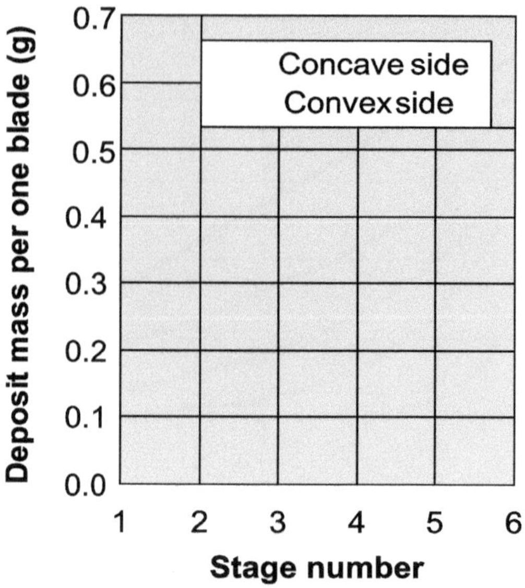

Fig. A.25 Measured distribution of deposits on axial compressor rotor blades in a two-shaft frame 5 gas turbine (Tarabrin et al., 1998).

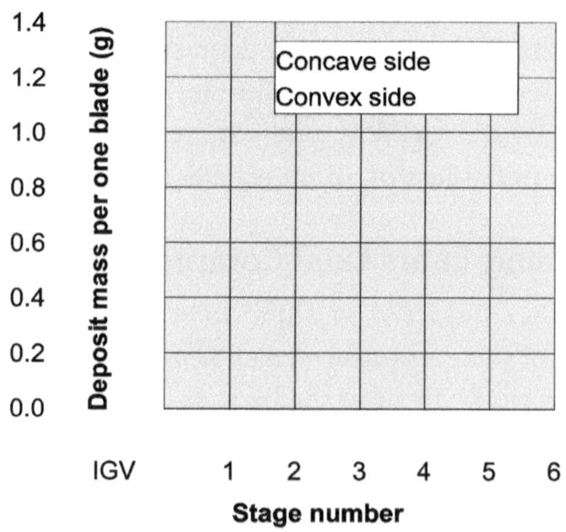

Fig. A.26 Measured distribution of deposits on axial compressor stator blades in a two-shaft frame 5 gas turbine (Tarabrin et al., 1998).

between compressor efficiency and mass flow rate change follows the following rules according to Zaba (1980).

- Percent change in mass flow is approximately equal to the percent change in compressor efficiency if all the stages are equally fouled.

- Percent change in mass flow is greater than percent change in compressor efficiency if the early stages are fouled (this situation is most commonly found in practice). Typically, the change in mass flow rate in percent equals 1.25 times the change in compressor efficiency in percent. For example if mass flow changes 2%, the compressor efficiency changes by ~0.8%.

Engine Configuration and Number of Shafts

Analytical studies by Tarabrin et al. (1998) examined the susceptibility of fouling based on the engine configuration. Three configurations were examined:

- Single-shaft gas turbines
- Two-shaft gas turbine, i.e., a gas generator and a free power turbine
- Three-shaft configuration with a two-shaft compressor (LP and HP) driven by their respective HP and LP turbines and a free power turbine

For the comparative study the same thermodynamic cycle parameters of pressure ratio, airflow, and ISF were used. The study examined the effect of a 1% drop in compressor efficiency with results being presented in Table A.2 and graphically in Fig. A.27. The table and figure indicate changes in power (δN_e), gas thermal efficiency ($\delta \eta_e$), airflow rate (δG), pressure ratio ($\delta \Pi$), and compressor speed (δN, δN_{LPC} δN_{HPC}).

For the case of constant turbine inlet temperature T_3 it can be seen that the single-shaft gas turbine exhibits an output drop of 2.82% compared to a drop in power for the three-shaft gas turbine of 4.32%. The physical

Table A.2 Coefficients of Influence of Gas Turbines With a 1% Drop in Compressor Efficiency (Tarabrin et al., 1998)

GT Unit Type	Nominal Regime Parameters	Conditions of Calculation	δN_e	$\delta \eta_e$	δT_3	δG	$\delta \pi$	δn	δn_{LPC}	δn_{HPC}
Single shaft	$T_3 = 1223\,K$ $\pi = 12.5$	n = const T_3 = const	−2.82	−1.41	0	−1.25	−1.25	0	–	–
Two shaft	$T_3 = 1223\,K$	T_3 = const	−3.5	−1.55	0	−2.07	−2.07	−1.4	–	–
	$\pi = 12.5$	n = const T_3 = const δF = varia	−2.82	−1.42	0	−1.25	−1.25	0	–	–
Three shaft	$T_3 = 1223\,K$ $\pi = 12.5$	T_3 = const	−4.32	−2.16	0	−2.16	−2.16	–	−1.16	−0.58

Note: Coefficients of influence are given in % for −1% of the axial compressor efficiency change.
δN_e = change in gas turbine unit output, $\delta \eta_e$ = change in thermal efficiency,
δG = change in airflow, $\delta \pi$ = change in pressure ratio; δn = change in compressor speed.
T_3 = turbine inlet temperature; F = power turbine variable nozzle area.

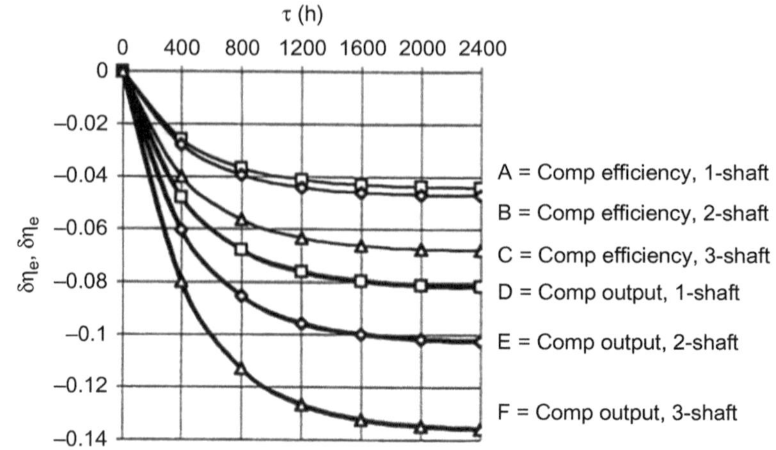

Fig. A.27 Influence of axial compressor fouling on efficiency and power for different shaft configurations (Tarabrin, et al. 1998).

explanation of this behavior is that the power match point is of a gas generator compressor, and a gas generator turbine is achieved by a greater change in air mass flow in comparison with the single-shaft gas turbine. The two-shaft turbine lies between the three-shaft and single-shaft turbines in terms of power drop (power drop of 3.5%).

In the case of a two-shaft turbine with variable power turbine nozzles (as may be found on mechanical drive split shaft gas turbines) the compressor speed can be kept close to constant by varying the power turbine area, thus resulting in a power drop of 2.82%.

Simulation Runs With Compressor Degradation on Six Gas Turbines

GTPRO simulation runs assuming an ambient temperature of 15°C by imposing compressor degradation were conducted on the following available gas turbines ranging in power from 38 to 174MW.

- Advanced technology high output gas turbine, single-shaft configuration (designated A)
- Moderate firing temperature gas turbines, single-shaft configuration (designated as B and C, with B being a scaled up version of C)
- Moderate firing temperature gas turbine, single shaft approximately half the power of C (designated D)

- Aeroderivative gas turbine with 2 compressor spools and a power turbine (designated E)
- Aeroderivative gas turbine with three compressor spools (designated F)

As before, the degradation steps imposed on these gas turbines were

- Step 1: New and clean, mass flow drop = 0%, comp. efficiency drop = 0%
- Step 2: Mass flow drop = 1%, comp. efficiency drop = 0.833%
- Step 3: Mass flow drop = 2%, comp. efficiency drop = 1.67%
- Step 4: Mass flow drop = 3%, comp. efficiency drop = 2.5%
- Step 5: Mass flow drop = 4%, comp. efficiency drop = 3.33%
- Step 6: Mass flow drop = 5%, comp. efficiency drop = 4.167%
- Step 7: Mass flow drop = 6%, comp. efficiency drop = 5%

The following assumptions were made for all simulations:

- Fuel CH_4, LHV = 50,047 kJ/kg supplied at 25°C
- Gas turbines run at 100% rating, inferred TIT control mode, control curve limited
- Site ambient conditions 1.013 bar, 15°C, and 60% relative humidity
- Inlet and outlet losses 10 and 12.45 mbar, respectively

In order to compare the gas turbine and their response to fouling, the simulation results were normalized where "1" represents the output and efficiency at Step 1 (i.e., new and clean condition with no deterioration imposed).

The results in terms of normalized power and normalized efficiency are shown in Figs. A.28 and A.29 and seem to corroborate Tarabrin's findings that the multispool machines tend to be more sensitive to fouling.

It is important to note that this analysis addresses the impact on performance when simulated fouling is imposed uniformly on all six gas turbines (i.e., the sensitivity). It does not address the susceptibility to fouling of these engines.

Research conducted by Arnulfi and Massardo (1993) states that the distribution of pressure rise in the compressor is also an important determinant in fouling deterioration behavior.

Site Location and Environment

The geographical area, climatic conditions, and the geology of the plant location and its surrounding environment are major factors that influence compressor fouling. Areas can be classified as desert, tropical, rural, artic, offshore, and marine. Salient data for different climatic conditions are provided in Table A.3.

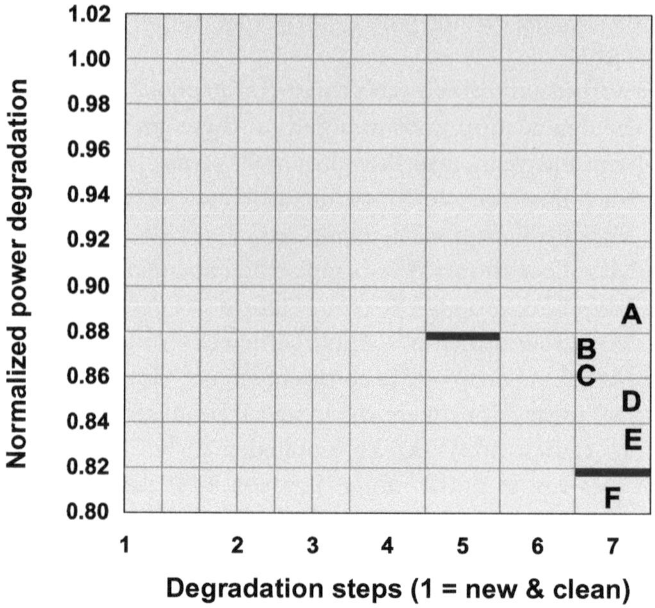

Fig. A.28 Normalized power output changes with fouling degradation steps for different gas turbine models.

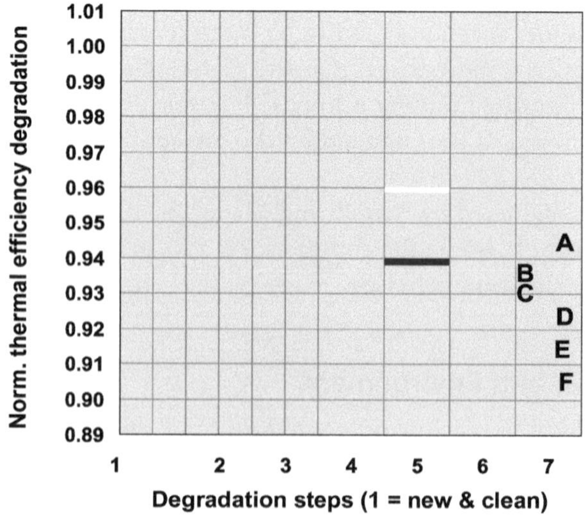

Fig. A.29 Normalized efficiency changes with fouling degradation steps for different gas turbine models.

Table A.3 Comparison of Operating Environments

Environment	Rural	Coastal and Offshore	Urban	Industrial
Type of dust/ foulant	Dry, usually nonerosive Possible agro-chemicals	Dry, nonerosive salt particles Corrosive mists Drilling mud etc.	Sooty/oily Can be erosive and corrosive	Sooty/oil Often erosive and corrosive Chemicals
Dust conc. (mg/m^3)	0.01–0.1	0.01–0.1	0.01–10	0.1–10
Typical particle size (μm)	0.01–10	0.1–3 salt <5	0.1–20	0.1–50
Effect on GT	Fouling (often minimal)	Fouling and corrosion	Fouling, and sometimes corrosion	Fouling, erosion and corrosion, due to acidic conditions
Temp. range (°C)	−20 to +30	−20 to +25	−20 to +35	−20 to +35
Weather conditions	Full range (dry, sunny, rain, fog, snow)	Full range, plus sea mist and freezing fog	Full range, plus hail and smog	Full range, plus hail and smog

Environment	Desert	Tropical	Arctic
Type of dust/foulant	Dry and erosive in sand-storm areas Fine, "talc-like" in dusty terrain	Nonerosive, but may cause fouling Pollen and insects	Nonerosive
Dust conc. (mg/m^3)	0.1–700	up to 135	0.01–0.25
Typical particle size (μm)	1–700	0.1–10	0.1–10
Effect on GT	Erosion Filter plugging with sand or insect swarms	Fouling	Plugging of air intake system with snow and ice
Temp. range (°C)	−5 to +45	+5 to +45	−40 to +5
Weather conditions	Long, dry, sunny spells; cold nights; high winds; sand and dust storms; some rain	High humidity, rain, insect, and mosquito swarms	Heavy snow, icing, high winds, possible insect swarms in summer

Airborne contaminants (dust, aerosols) and their type (salts, heavy metals, etc.), their concentration, particle sizes, and weight distribution as well as the vegetation cycles are important parameters influencing the rate and type of compressor fouling.

Plant Layout and Design

The rate of compressor fouling deterioration and foulant type are strongly affected by predominant wind directions.

Factors that must be taken into account include

- Orientation and elevation of air inlet ducts
- Location of water cooling towers or air to air heat exchangers
- Possibility of gas turbine exhaust recirculation into the air inlet
- Orientation of lube oil breathers and vapor extractors
- Location of highways, and adjacent plants, that can contribute to the fouling problem

CFD predictions can be used to understand the phenomenon especially in grass roots installations. Other important plant design parameters, which affect the rate of compressor fouling, are the selection of appropriate air inlet filtration systems (self-cleaning, depth loading, etc.), the selection of filter media, the number of filtration stages, weather louvers, inertial separators, and mist eliminators. Design parameters such as the air face velocity through the filters, filter loading, and their behavior under high humidity and high pressure drops are also critical.

Plant Maintenance Philosophy

The quality and maintenance philosophy of an operating plant will have a significant effect on the fouling deterioration rate. The following plant maintenance activities will have an impact:

- Quality of maintenance of the air filter system. This includes checking the media quality, monitoring the differential pressure, taking maintenance action to avoid leakages and ensuring proper storage of the spare air filter elements to avoid damage.
- If an evaporative cooling system is present the degree of maintenance becomes even more important. Evaporative media must be checked for signs of deterioration or separation from the steel framework or loss of integrity of the mist eliminator. Unequal wetting of the media can create droplet carryover. With fogging systems, constant checks on the quality of the demineralized water must be made.
- Method of detection of fouling deterioration. This could include some form of performance monitoring or tracking of key aerothermal parameters.
- Schedule and frequency of compressor washing
- Maintenance of the gas turbine bearings and seals to prevent oil leakages into the inlet plenum.

In general a plant that is conscious of the fouling problem and the losses associated with it will adopt a maintenance schedule to manage and mitigate the problem. Unfortunately, in several mechanical drive applications the power margins that are designed into the system (excess gas turbine power over compressor absorbed power) may promote a negligent attitude toward fouling control.

Climatic Parameters

Ambient temperature, relative humidity, and climatic conditions such as fog and smog will strongly impact compressor fouling. The change in differential pressure in an inlet system over a day is shown in Fig. A.30, with the effect of fog (increase in differential pressure) being evident during the early hours of the day.

In some cases, turbines have been known to trip due to excessive backpressure caused due to morning fogs. In other cases, excessive sandstorms have also been known to create high differential pressure conditions.

Correlation Between Rate of Power Degradation and Atmospheric Condition

Based on various field tests and observations (Stalder, 1998), a correlation between rate of power degradation and atmospheric conditions prevailing at site has been established. Stalder's studies included extensive testing

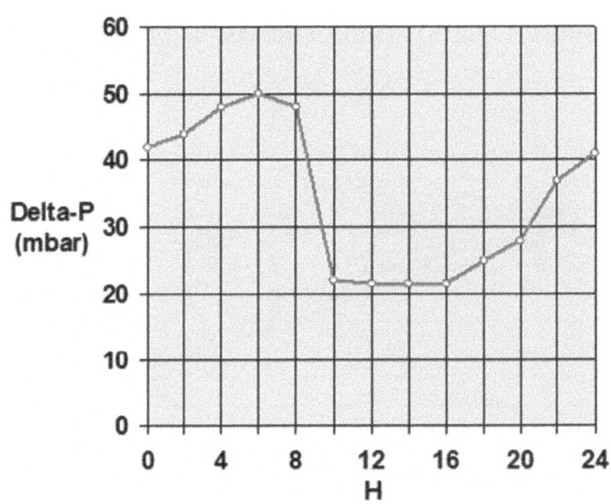

Fig. A.30 Filter differential pressure changes with time due to fog in early morning.

covering 18 months and utilizing several types of wash cycles including on-line and crank washing. The salient results are presented below.

Out of 40 measured continuous operating periods (without shutdowns and start-ups), a total of 14 operating periods each between 70 and 72 h can be directly compared in Fig. A.31 This graph shows the power output measurements were made at the beginning of each period after on-line washing (100% reference point) and at the end of each period, prior to on-line washing of the next period. One can see that there is a large spread in the power losses over such a short period, and the highest loss in performance was 3.1%. Interestingly, the output at the end of one period shows a gain of 0.5% power output. This surprising result was obtained on the same unit, with the same air inlet filtration system, the same washing nozzle system, the same washing procedure, and the same detergent.

It is generally assumed that power losses will depend on the amount of humidity in a specific environment. With the data collected during the comparative test periods above, the total quantity of water and vapor mass flow ingested by the compressor was determined. Stalder calculated the compressor air mass flows by means of a heat balance. The average ingested total humidity (water and vapor) amounted to 7.7 tonnes/h, i.e., a total of 548 tonnes during 70 operating hours. The lowest average value during a period was 4.1 tonnes/h and the highest was 11 tonnes/h.

The distribution in Fig. A.32 shows the measured power losses versus the total quantity of humidity (water and vapor) ingested by the compressor for each of the selected comparative 70-h operating periods. The results of this test clearly indicated a strong correlation between the mass flow of absolute humidity and output deterioration. The loss in power output on the unit

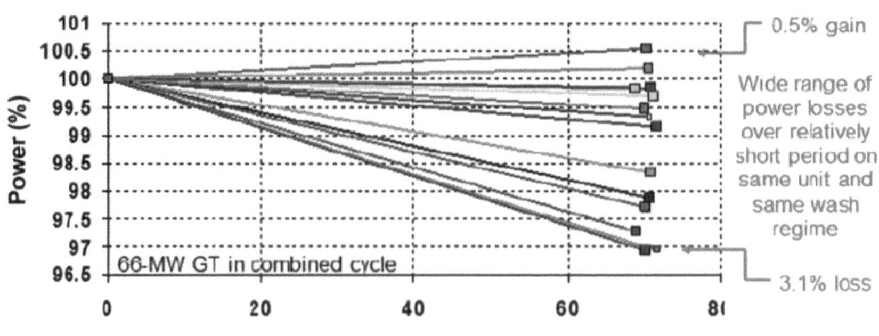

Fig. A.31 Power losses over 70 continuous operating hours.

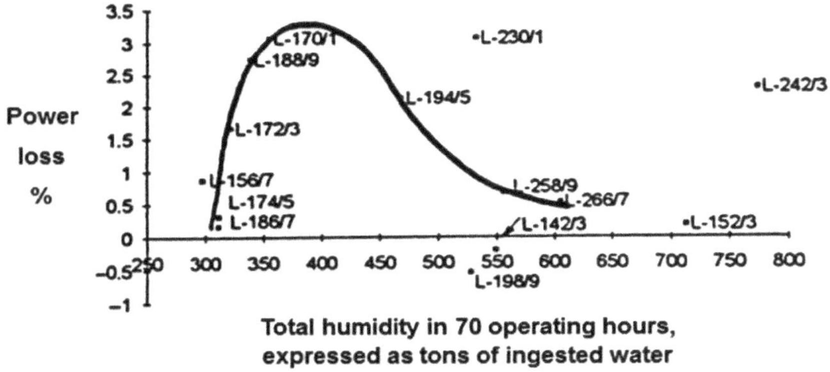

Fig. A.32 Correlation between total absolute humidity and power loss over 14 comparative operating periods (Stalder, 1998).

tested increases with increasing mass flow of absolute humidity until it reached a peak at ~400–450 tonnes (total over 70 h) before decreasing again.

Correlation Between Power Loss Gradients and Humidity

Due to a drop in static pressure at the bellmouth caused by the acceleration of the air near the IGVs, the humidity in the air will start to condense. To give an idea of the amount of water condensed, Stalder (1998) provides the following example for a 250 MW gas turbine with an airflow rate of 500 m3/s operating at an inlet Mach number of 0.5 at the IGVs:

- At 12°C gas turbine compressor inlet temperature and 90% relative humidity (RH), the total condensing water mass forming droplets will be up to 6.3 tonnes/h.
- At 12°C and 60% RH, the total condensing water mass forming droplets will be up to 3.1 tonnes/h.
- The latent heat released by the condensing water (i.e., increasing the compressor inlet temperature) will be higher at 90% RH. Therefore the final static temperature drop in the air inlet at 90% RH is 7°C whereas it is 10°C at 60% RH.

Summarizing Stalder's findings with respect to humidity:

- *Influence of humidity*: Surface wetness of compressor blades operating in saturated condition will modify the aerodynamic boundary layer promoting fouling and a subsequent drop in performance.
- *Latent heat release*: The ingested air temperature will increase at the compressor bellmouth entry as condensation occurs and latent heat is released, thus reducing cycle efficiency.

- *Humidity and water-wettable/-soluble deposits*: With lower amounts of condensed water droplets, ingested foulants will combine with the water droplets and deposit on the vanes and blades. The rate of deposition will increase with the resulting roughness. Above a certain amount of condensed water droplet mass flow that is formed, the blading will be naturally washed and power losses due to fouling will be recovered to some extent. (This explains the phenomenon of "self-cleaning" that several operators have experienced but do not understand.)
- *Humidity and hydrocarbon type of deposits*—a similar effect as that above will occur for a small amount of condensed water droplet mass flow. However, as water is a poor hydrocarbon solvent, the natural washing effect as the condensed water droplet mass flow increases will be very limited. Power losses due to fouling will continue until equilibrium is reached.
- *Humidity and combination of water-wettable/-soluble and hydrocarbon type of deposits*: This type of deposit is very common, and depending upon the mass relationship of hydrocarbon vs water-wettable/-soluble foulants, and their respective embedment in the deposit layers, the "natural washing" effect when condensation is high can also be limited.
- *Fouling rate at low ambient temperatures*: The flattening of the saturation curve in the low ambient temperature range (below about −10°C, 14oF) in the psychrometric chart shows that the amount of water droplets which can combine with foulants is significantly reduced, resulting in lower fouling.
- *Duration and sequence of operation in a saturated condition*: The changes in rate of power losses noticed over a given operating period will be strongly influenced by the duration and sequences of operation under saturated high humidity conditions (vs a dryer condition).

Humidity Effects on the Fouling of Axial Compressors

As air passes through the intake and filtration system, it proceeds at a very low velocity with filter face velocities being typically around 3 m/s. As it approaches the compressor face, the air accelerates to a high velocity (0.5–0.8 Mach number). This results in a static temperature reduction of about 10–15°C. The saturation air temperature also drops. If the relative humidity is high enough, it is possible that the static air temperature falls below the saturation air temperature. This causes condensation of water vapor, which is a common occurrence in most gas turbines when ambient relative humidity is high.

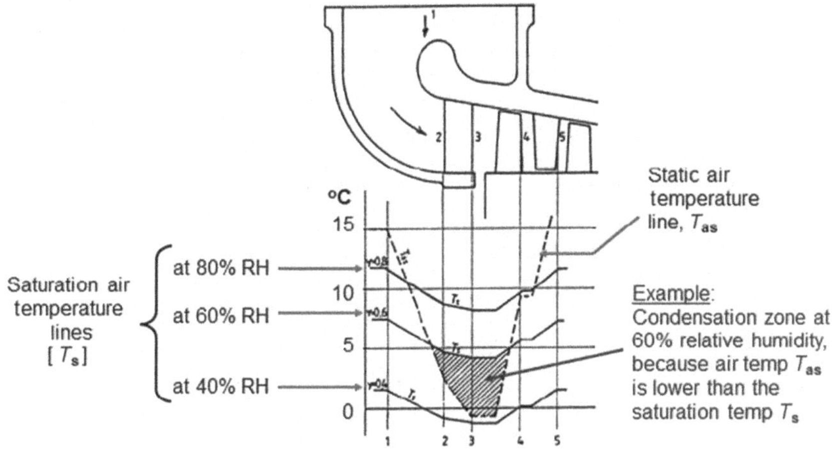

Fig. A.33 Intake temperature depression and condensation (Zaba and Lombardi, 1984).

The process is depicted in Fig. A.33 (Zaba and Lombardi, 1984). In this figure, T_{as} is the static air temperature, T_s is the saturation air temperature, and γ is the relative humidity. Three values of relative humidity are shown (40%, 60%, and 80%). As can be seen in this figure, when the relative humidity is 60%, the air temperature T_{as} is lower than the saturation temperature (in the shaded region of the figure), resulting in condensation of the water vapor.

Filters tend to unload salt (leeching effect) under high ambient humidity conditions and this is a factor that is often neglected. It is this factor that causes the sudden fouling of compressors during periods of ambient fog.

Particles then form nuclei for the water droplets and start to adhere to the blading. As the air progresses to the rear compressor stages, it gets hotter and drier and typically causes less fouling in the rear stages. A photograph of salt leeching through a filter is shown in Fig. A.34.

Stalder and Sire (2001) have conducted detailed analytical and experimental work on salt percolation through gas turbine air filters and some of the salient observations and results are presented below.

- The ingestion of airborne sea salt through a filter can result in high temperature corrosion damage. With the high temperature engines available today this is of significant importance as turbine manufacturers limit the total amount of salt entering the engine to low values such as 0.2–0.5 ppmw. Total contaminants that enter the engine must be calculated based on airflow, water injection flow, and fuel flow. The total contaminant entering the engine is given by

Fig. A.34 Salt leeching through intake filter (Cleaver, 1990).

$$\text{TCL} = I_f + [I_{air} x A/F] + [I_w x W/F] + [I_S x S/F]$$

where TCL is the total contaminant level, ppmw; I_f is the contaminant level in the fuel, ppmw; I_{air} is the contaminant level in the air, ppmw; I_w is the contaminant level in the injection water, ppmw; I_{stm} is the contaminant level in the injection steam, ppmw; A/F is the air to fuel ratio for the gas turbine; S/F is the steam to fuel ratio; and W/F is the water to fuel ratio.

- With regard to limits of airborne contaminants, the following example is provided: Natural gas contaminated with salt at a sodium concentration of 0.1 ppm and airborne salt contamination of 5 ppb will result in an equivalent fuel Na content of 0.35 ppm—assuming an air to fuel ratio of 50. Specific OEM requirements should be considered, as this is dependent on blade metallurgy, coating technology, and cooling approach. The performance of filters in limiting ingress of salt is therefore extremely important.
- Studies performed on used filters indicate that percolation of salt laden water can occur in both surface and depth loading type of filters. The mechanism is supported by the presence of dissolved acid gases (SO_x and NO_x) found in industrial environments.
- The large area of the air filter media may also act as a reaction site for various chemical processes and reactions. The real challenge is to

Table A.4 Analysis of Air Filter Element After 15,000 Hours Operation in a Coastal and Industrial Environment

Concentration (μg/cm²)	**Front Location "Upstream" Dirty Air Side**	**Middle Location "Middle"**	**Bottom Location "Downstream" Clean Air Side**	**Average of the 3 Sample Locations**
Soluble fractions: $Na^+ + K^+$ cations	12	66	158	79
Total cations$^{(+)}$	80	402	683	388
Chloride anions, Cl^-	18	201	213	144
Total anions$^{(-)}$	161	875	1350	795
Total soluble fraction	241	1277	2033	1183
Total insoluble fraction	779	656	609	681

Sampling locations represent ~1/3rd sections along cylindrical air filter element. Results are expressed in micrograms per sq. cm. of filer element surface area.

determine the efficiency of the media under real-life operational conditions where aggressive gases may exist in the environment.

- Optimal selection of filter media is of supreme importance and the system must be tailored to climatic conditions.
- Good filtration must be coupled with a carefully planned on-line and off-line washing regime to minimize salt deposits on compressor airfoils.

Stalder and Sire (2001) present results of a very interesting investigation into salt ingress into the combustion air by the means of examining filters. Details of the analysis of one of the filter elements are shown in Table A.4. The results show that higher concentrations of chlorides and sodium are found *within* the filter near the clean air side, clearly indicating that percolation or leeching has taken place. These analysis results are for a cylindrical type air filter element after 15,000 h of operation in a coastal and industrial environment. Additional details on airborne contaminants are provided in Hsu (1988).

Inlet Air Filtration

There are several types of filters that can be categorized in the following groups:

- *Inertial filters*: The objective here is to make the air change direction rapidly causing separation of dust particles. These filters are permanently fixed and require minimal maintenance. Inertial filters typically operate at face velocities of 20 ft/s.

- *Prefilters*: These are medium efficiency filters made of cotton fabric or spun fiberglass. They are relatively inexpensive and serve as "protection" for high-efficiency filters.
- *Coalescers*: These are constructed by the use of wire mesh that acts as an agglomerator. The mist in the inlet air is agglomerated and the moisture is thus removed. (Other vane type designs are also available.)
- *Louvers and vanes*: These are typically used in the first stages along with coalescing filters to remove water droplets.
- *High-efficiency filters*: These filters remove smaller particles of dirt. They are typically barrier or bag type filters.
- *Self-cleaning filters*: These consist of a bank of high-efficiency media filters, and air is drawn through the media at a low velocity. At a predetermined pressure drop of about 4.98–7.5 mbar (2–3 in. WG) a reverse blast of air is used to remove dust buildup.

Goulding et al. (1990) have provided a detailed treatment of the technical considerations in the selection of gas turbine filters. Details of offshore platform filtration systems are provided in Kimm and Langlands (1985) and Schweiger (1983). Zaba and Lombardi (1984) have detailed experiences with different types of filters.

Practical Aspects Relating to Air Filter System Integrity

Air tightness is a must for any gas turbine inlet system as even the most efficient filtration system will be useless if unfiltered airflow leaks in and enters the compressor. Some common causes of leakage are

- Bypass door leakage
- Poor gaskets and seals at flanges
- Modifications made on the inlet ducting. Over the years, personnel may add structures or devices to the inlet system that may cause problems

Some important considerations in intake filter design are

- Aerodynamic design should be such as to keep intake velocities uniform across the entire filter area.
- The filter housing should be of a bolted and welded design fabricated of steel no <4.76 mm (3/16″) thick and reinforced by steel members. The filter house should withstand 30 mbar (12 in. WG) pressure. All seams and joints should be airtight. All nuts and bolts used inside the clean air plenum should be welded after assembly to prevent air leaks and foreign object damage to the turbine.

- Design should facilitate change-out of all filters from the upstream side. Filter change should be possible without turbine shutdown. Filter elements should be designed for quick change-out, avoidance of blind assembly and loose retaining nuts, ungasketed washers, etc.
- The filter design should ensure that the inlet air is drawn at least 3 m (10 ft) above grade level. In some locations a greater height may be required.
- A stainless steel trash screen with 25.4 mm (1 in.) square mesh should be provided in the transition section between the clean air plenum and the compressor intake.
- Avoid the use of gravity weighted bypass doors. Bypass doors are designed to permit emergency airflow to the engine when intake pressure drop rises above a critical value. Bypass doors are typically gravity operated or power operated. The gravity type has earned a reputation for unreliability. Poor sealing, hinge corrosion, and improper operation have made the bypass door a weak link in inlet filter design.
- All filter seal points should be reviewed during the design phase. Poor intake sealing has allowed leaks through bypass doors, access doors, and flanges on the intake filter. In several situations, flange distortion has allowed air ingress. Users should specify types of seals required and call for a filter house integrity test under specified depression to ensure air tightness.
- System design in the case of pulsed cleaning systems should be such as to minimize flow distortions and pressure pulsations due to pulse cleaning. More than 5% of the total filter elements should not be cleaned simultaneously.
- Filter system pressure drop is an important parameter affecting gas turbine performance. It is important to consider both new filter pressure drop and the pressure drop increase over time.

Filter House Design and Corrosion

Experience has shown that stainless steel (316 L) construction is *by far the most sensible approach* especially in offshore and coastal areas. The life-cycle costs when one considers the constant painting that has to be done to carbon steel filter systems, is several times the small incremental first cost increase for the use of SS 316 L construction. It is indeed unfortunate that the desire to minimize first cost often results in filter systems that cause great harm to the

Fig. A.35 Corroded floor of inlet duct (Cleaver, 1988).

project in term of gas turbine deterioration, and even subsequent damage that can be caused by severe rusting and other distress over the project life cycle.

A highly corroded floor of an inlet filter is shown in Fig. A.35.

Corrosion of inlet systems is widespread especially when they are made of carbon steel. Corrosion of the filter house downstream of the filters totally defeats the purpose of having air filtration. Painted carbon steel structures need constant painting and often, poor application of paint and other problems involving dissimilar materials start the corrosion process only weeks after the painting is completed. Proper painting requires the use of shot blasting to fully clean the parent surface prior to the painting process. Modern technology has allowed the use of high performance SS 316L systems that include relatively light elements, and which are very effective against driving rain and high humidity. These filter systems use special high velocity vane separators that have proven very effective for offshore applications.

The choice of material is also very important as SS 304 or 321 steels do not have sufficient corrosion protection. The presence of chromium (up to 18.5%) in SS 316L helps to build a passive protective film of oxide and prevents corrosion. Along with the 10%–14% nickel content, 316L can be easily welded. In offshore and hostile environments, it is best to have all the system components (including vane separators and door hinges) made of 316L stainless steel. Proper attention must also be paid to drain systems to enable water to be removed from the system. An excellent treatment of air filtration is made by Cleaver (1990). Details regarding the testing of gas turbine air filters are provided in Gidley et al. (1993).

Filter System Pressure Drop

The effect of high pressure drop in the inlet filter system is a reduction in the inlet density and mass flow of the gas turbine and a reduction in the turbine section expansion ratio. The effect of inlet system pressure drop is provided for a Frame 5 gas turbine in Table A.5. Salient cycle conditions for the engine are provided for inlet pressure drops of 0 through 12.54 mbar (0–5 in. W.G). As the inlet system differential pressure increases, the pressure to the turbine section drops and results in a lower expansion ratio and an increase in the EGT. A power drop of 0.4% per inch of WG differential pressure and a heat rate increase of 0.19% per inch of WG differential pressure emphasize the importance of this parameter.

Table A.5 Impact of Inlet Filter Differential Pressure on a Frame 5 Gas Turbine
Inlet Differential Pressure Sensitivity Study Simulation

Fuel = CH4, supplied @ 77 F, LHV = 21,517.58 BTU/lb
G.T. @100% rating, inferred TIT control model, CC limit
Site ambient conditions: 14.7 psia, 100 F, 60% RH

GE 5371PA	Case 1	Case 2	Case 3	Case 4	Case 5
INLET DELTA P, inch WG	0	1	2	3	4
Computation Result, Thermoflow-STQUIK	OK	OK	OK	OK	OK
GT gross power (kW)	21,861	21,773	21,685	21,598	21,510
GT gross LHV efficiency (%)	26.84	26.80	26.75	26.70	26.65
GT gross heat rate (BTU/kWh)	12,711	12,734	12,758	12,781	12,805
Compressor inlet mass flow (lb/s)	245.0	244.4	243.7	243.1	242.5
Compressor inlet temperature (F)	99.86	99.86	99.86	99.86	99.86
Turbine inlet mass flow (lb/s)	238.3	237.8	237.2	236.6	236.0
Turbine inlet temperature (F)	1761	1761	1761	1761	1761
Turbine exhaust mass flow (lb/s)	248.5	247.9	247.3	246.7	246.1
Turbine exhaust temperature (F)	939.6	940.3	941.0	941.8	942.5
GT fuel HHV input (kBtu/h)	308,341	307,663	306,985	306,306	305,627
GT fuel LHV input (kBtu/h)	277,881	277,270	276,659	276,048	275,436
GT fuel flow (lb/s)	3.587	3.579	3.571	3.564	3.556
% Change in Power, per 1″ WG	0.40				
% Change in HR, per 1″ WG	−0.19				

Effects of Fouling on Gas Turbine Operation and Maintenance

Aerothermodynamic Effects of Compressor Fouling

The observable effect of compressor fouling is a drop in thermal efficiency (increase in heat rate) and a drop in output. The axial flow compressor is a sensitive component that requires smooth aerodynamic surfaces. Fouling causes an alteration in the shape and profile of the blading (increased surface blade roughness), and this reduces air flow rate, pressure ratio (of the overall compressor), and compressor efficiency. Modeling details may be found in Lakshminarasimha et al. (1994) and Tabakoff (1988).

Fig. A.36 shows the changes in discharge pressure due to fouling at different ambient temperatures for a heavy-duty gas turbine.

Surface roughness caused by compressor fouling increases the profile losses of the blading. These losses appear as a boundary layer momentum thickness, which increases with increasing roughness in the blade profile. In simple terms, the drag increases and results in an increase of the specific work (kW/unit mass flow) of the compressor.

For example, for the simulated values of the 40 MW class gas turbines modeled with degradation steps in Figs. A.11–A.13: The specific compressor work increases from 158.77 kW/lb/s in the undeteriorated condition to 163.82 kW/lb/s in the maximum deterioration case.

A detailed treatment of losses is provided in Koch and Smith (1976). A treatment of roughness effects on the aerodynamics is provided in Bammert

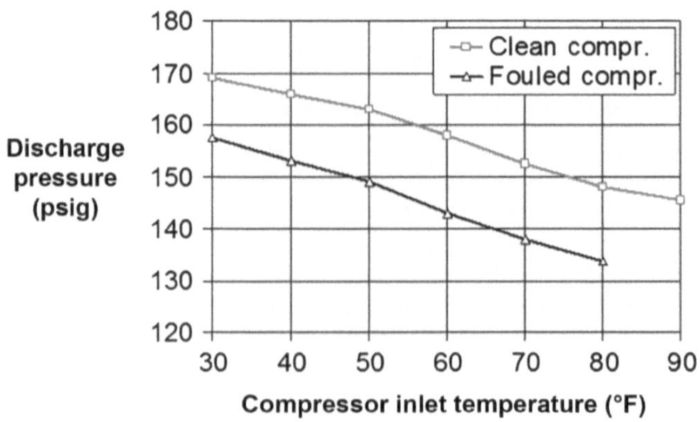

Fig. A.36 Drop in compressor discharge pressure before and after fouling.

and Woelk (1980). An interesting treatment of the effect of Reynolds number and blade surface roughness is provided in Schaffler (1980).

Fouling, Airflow Distortion, and Compressor Surge

As fouling reduces the mass flow (flow coefficient) in the first compressor stage, the performance of the latter stages is affected and the operating point on the first-stage characteristic moves toward the left. The first-stage pressure ratio is thus increased. This causes a higher density at the inlet to the second stage. Thus there will be a further reduction in second-stage flow coefficient. This effect progresses through successive stages until stall occurs in a rear stage and surge is triggered. Details on the effects of stage characteristics on axial flow compressor performance may be found in Stone (1958).

Dundas (1986) has conducted a detailed analytical investigation into the deterioration of turbine operation including drop in compressor efficiency, fouling, first-stage nozzle distortion, internal bleed seal deterioration, drop in turbine efficiency, inlet filter fouling, and low fuel heating value. These parameters were examined to study the effect on the turbine operating line. His study concluded that compressor fouling had a pronounced effect on the operating line.

Whereas this effect causes the movement of the operating line toward the surge line, there are other factors that can cause movement of the surge line itself. Erosion of compressor blading can affect boundary layer development and increase the tendency toward separation. Stall can therefore occur at a lower incidence angle than with smooth compressor blading. Heavy erosion can also reduce blade tip chords, thereby reducing blade tip solidity, which would adversely affect stage stability.

The importance of considering fouling effects on surge becomes more important with the use of gas turbines in combined cycle cogeneration applications and with IGV control and steam injection applications. Closure of the IGVs during part load operation restricts airflow and this, in conjunction with severe fouling, can promote surge. Steam injection results in a higher back pressure on the compressor, thus moving the operating point closer to the surge line.

There have been several cases where excessive distortion of the inlet airflow has triggered a surge event resulting in compressor damage. Icing, for example, causing uneven inlet circumferential distortion, or uneven clogging of filters (possibly due to a bend in the inlet duct before the filter or improper inlet system design), can create distortion effects that could result in surge. Studies and results relating to the flow in inlet ducts are given by Manfrida et al. (1988).

Effects of Fouling on Compressor Blading Integrity

While fouling cannot be said to be a major cause of compressor blade failure, it can contribute to blade problems as indicated below:

- By promoting surge or rotating stall which may have a dangerous effect on blades.
- In some cases, blading natural frequencies can be affected by the increase in mass due to dirt buildup on the blading.

Blade roughness, and therefore efficiency and performance, can be adversely affected by corrosion, erosion and fouling. Excessive dirt on the blades can cause imbalance, and a consequent increase in running speed vibration. In some cases dirt can get between the bearing surfaces of the blade root, causing the blades to operate in an abnormal position, which adds to the stresses. If the root constraint is changed due to buildup in the fir tree region, a change in natural frequency could result (as the boundary condition changes). The blocking (or partial blockage) of cooling passages of hot section stators and blades can be caused by fine foulants (typically $<5\,\mu m$). As the cooling air is bled from the compressor, foulants can enter the cooling system. Cement dust, coal dust, and fly ash can be responsible for this problem. The effects can be improper cooling and accelerated thermal fatigue, though typically the effects are gradual in nature.

- Foulant buildup on compressor blading can lead to a serious corrosion problem, especially when humidity is high. Airborne salt may be ingested as dry particles or as droplets of brine, and moisture will be absorbed during high humidity operating conditions or during water washing. Salts, mineral acids, and aggressive gases (e.g., SO_x, NO_x, Cl_2) along with water can cause pitting of compressor blades due to electrochemical corrosion mechanisms. This can lead to local stress raisers that can diminish blade fatigue life. Compressor coatings are of value here.
- 5–10 μm particle size represents the transition zone between fouling and erosion, and particles causing erosion are normally 10 μm or greater (note: $10\,\mu m = 1/15$ diameter of human hair). Erosion impairs aerodynamic performance and can affect the mechanical strength of the blade.

Erosion first increases blade surface roughness, thus lowering efficiency slightly. As erosion progresses, airfoil contour changes occur at the leading and trailing edge as well at the blade tip. Severe erosion has also been known to cause changes in blade natural frequency.

- On relatively small gas turbines, a 0.1 mm coating applied to the blading can cause a flow reduction of 10% and a reduction of compressor efficiency of 5%.

Foreign Object Damage (FOD)

Although not linked directly to fouling, this subject is mentioned because it can be caused by a loss in filter integrity. Damage is typically caused to the forward compressor stages, although in some cases the foreign object also works its way to rear stages and causes damage. Damage is a function of foreign object size and composition, blade construction and impact location. It can lead to direct or secondary failure. Foreign object damage can be caused by ice, failed intake section components, or by materials and tools left in the inlet plenum.

Compressor Section Corrosion

Experience has shown that deposits on compressor blades often contain sodium and potassium chlorides. These combine with water to form an aggressive solution that promotes pitting corrosion of the blades (typically a 12% chrome steel such as type 403 or 410). Water condenses due to the acceleration of the air prior to the IGV and the salt particles are dissolved and pass through the compressor. The water evaporates as it moves through the compressor and at times, salt is found deposited on compressor blading. In a salty environment, the fatigue strength of steel can drop 50–60%, and this situation is worsened when notches due to corrosion pitting are present. Even with effective air filtration, the conditions of fog, humidity, or rain can cause migration of the salt through the inlet filter (leeching) and into the compressor. Blading on which corrosion deposits have formed is shown in Fig. A.37.

Haskell (1989) states that corrosion is rarely observed beyond the 8th compressor stage, as no moisture will survive at the temperatures beyond this point. As shown in Table A.6, sulfur dioxide [SO_2] and hydrochloric acid [HCl] can create very acidic conditions on the compressor blading, even in the parts per billion concentration range. This situation is acute in heavy industrial areas or close to chemical manufacturing plants. Furthermore,

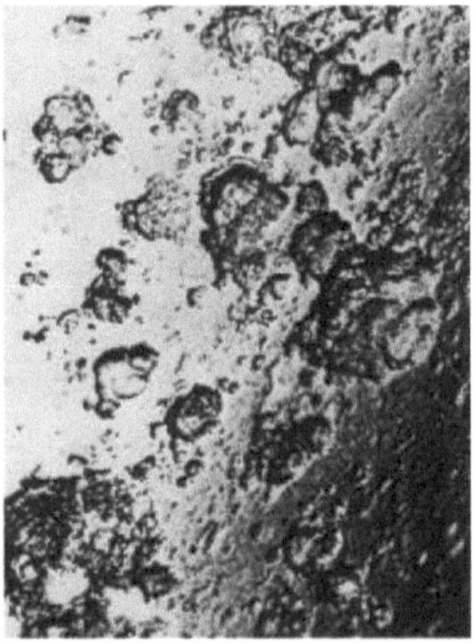

Fig. A.37 Pitting corrosion on surface of compressor blade.

Table A.6 Effect of Ambient Gases on Acidity at the Compressor Inlet

	Sulfurous acid	
Ambient SO_2 (ppb)	**Dissolved SO_2 (ppm)**	**pH**
1	0.2	5.5
10	0.64	5.0
100	2.0	4.5
1000	6.4	4.0
10,000	19.8	3.5
	Hydrochloric acid	
Ambient HCl (ppb)	**Dissolved HCl (ppm)**	**pH**
1	1600	1.44
10	5500	0.94
100	17,600	0.44

Based on Haskell, R.W., 1989. Gas turbine compressor operating environment and material evaluation. ASME International Gas Turbine and Aeroengine Congress, Toronto, Canada, June 4-8, 1989, ASME Paper No: 89-GT-42.

even moderate relative humidity conditions of about 50% will result in water formation at the compressor inlet due to the intake temperature depression. A detailed treatment of gas turbine blade failures including underlying causes and troubleshooting is provided in Meher-Homji and Gabriles (1998).

Detection of Fouling

Gas turbine manufacturers and operators usually develop guidelines to define when fouling deterioration calls for corrective action. This is based on a combination of load and exhaust gas temperatures (EGT). Users also monitor compressor discharge pressure and compressor efficiency. Graphs can be plotted to show expected (clean) versus measured parameters. However, it is the opinion of some operators that the only way to detect a fouled compressor is by visual inspection. Unfortunately, though, with most turbine designs this means shutting down the unit, removing the inlet plenum hatch, and visually inspecting the compressor inlet, bellmouth inlet guide vane (IGVs), and visible early stage blading.

The following factors can be used as indicators of fouling:

- Drop in power output for given ambient temperature and corrected speed
- Drop in compressor mass flow rate on fixed geometry engines
- Drop in compressor efficiency and pressure ratio (or discharge pressure)

The most sensitive parameter of the above factors is the mass flow rate.

The real problem is to detect fouling in time to prevent a significant power drop, and before a fuel penalty cost has been incurred. Several philosophies are in use. Some operators believe in regular periodic washing of the machine, whereas others base the washing requirement on a certain set of performance parameters. The philosophy utilized is a function of normally expected fouling levels, its severity, washing effectiveness, and plant operation criteria. Measurement of air-intake depression is a practical and economical method for fixed geometry machines. The technique involves measuring intake depression as an analog of airflow rate. In this approach, the gas turbine inlet bellmouth is utilized as a flow meter. This approach has been successfully used by Scott (1979, 1986).

By means of suitable software, data available in the gas turbine control system can often be used to monitor compressor deterioration. The set of graphs taken on a Frame 7EA gas turbine engine (Dusatko, 1995) show the correlations very clearly. In Fig. A.38, the general relationship between compressor discharge pressure (CDP) and compressor inlet temperature (CIT) can be seen. These data were taken over several hundred hours of operating data at full load conditions. The rule of thumb derived from this figure is that a 0.138 bar (2 psig) loss of compressor discharge pressure is equivalent to one gross MW of power for these particular Frame 7EA gas

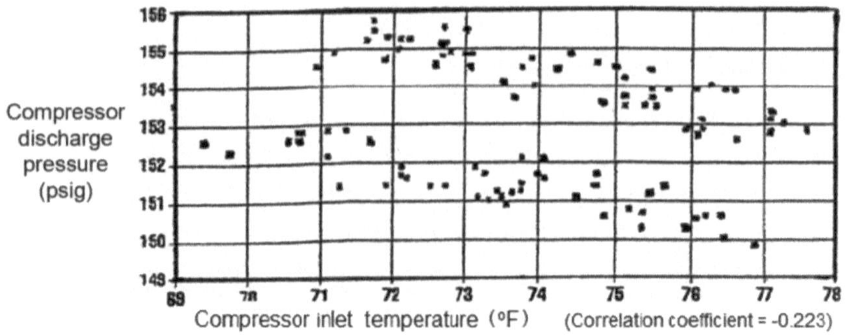

Fig. A.38 Scatter plot showing effect of crank wash on compressor discharge pressure vs compressor inlet temperature.

turbine engines. The effect of a crank wash on the CDP vs CIT plot is shown in this figure.

A common approach to trend gas turbine output is to use a power capacity factor defined as

$$\text{Power Capacity Factor} = \text{MW}_{\text{Actual}}/\text{MW}_{\text{Expected}}$$

The MW expected is the corrected "new and clean" performance test output, which is typically 3%–5% higher than the "guarantee" output.

Corrections should be made for

- Inlet temperature
- Inlet pressure
- Specific humidity
- NO_x water injection rates
- Inlet and outlet Pressure drops
- Speed corrections (this is not a major issue for power generation applications with single-shaft gas turbines unless significant off frequency operation is experienced. This happens in some countries due to grid "under-frequency" problems).

A baseline development process and the use of scatter plots for deterioration analysis of advanced F class turbines are provided in Meher-Homji et al. (1993).

Mapping of Compressor Performance to Monitor Deterioration

It is very valuable to monitor compressor performance by creating a compressor map and comparing the performance behavior to the predicted characteristics. Corrected speed and flow are the best correlating parameters as they allow variations in ambient temperature and pressure drop through an inlet filter to be incorporated on the map. The corrected speed represents the wheel tangential Mach number whereas the corrected flow represents the through-flow Mach number. Practical details of monitoring of compressor performance from a standpoint of detecting problems are provided in Dundas (1982) and Dundas et al. (1992).

Once a judicious schedule for on-line and off-line washing has been established, it is important to monitor the performance of the gas turbine and track for unexpected events and monitor the efficacy of washing. It is important that all parameters be monitored and appropriate corrections made using corrected speeds, temperatures, and flows that can then be trended. For aeroderivative engines the control approach must be taken into account (i.e., speed or temperature control). For aeroderivatives with multiple compressor spools, the interrelationship between the corrected LP and HP compressor spool speeds can provide a valuable indicator of compressor deterioration. Special performance monitoring considerations when operating at peak loads are provided in Syverud et al. (2003).

Control of Fouling by Compressor Washing

Fouling is best controlled by a combination of two methods. The first line of defense is to employ a high quality air filtration system. However, as fouling will inevitably occur, compressor washing should be used to control its impact.

Compressor washing is an area in which strong and divergent opinions exist. Washing efficacy is so site specific that approaches which work for one site may not be appropriate for another. Controversy is often caused by polarized opinions relating to wash procedures, wash media, and techniques. Some of the highlights are presented below in an attempt to present the overall picture and the practical field experiences of Stalder (1998).

Operators must determine the best approach for their gas turbines by trial and error in terms of wash technique, use of on-line washing, which cleaners should be used, and the frequency of washing. This is a complex technical–economical scenario, and also depends on the service that the gas turbines are in. For example, by not having the ability to shut down for crank washes, IPP operators and merchant power plants may need to be more aggressive in controlling fouling. A useful set of papers relating to compressor washing have been provided in Stalder (1998), Stalder and van Oosten (1994), and Bagshaw (1974).

Several different methods of gas turbine compressor cleaning have been applied over the years, but "wet cleaning" has been found to be by far the most effective and economic technique. However, today's sophisticated large industrial engines and blade coatings require appropriately designed cleaning systems to ensure operational safety, reliability, and optimum efficiency. Two different wet cleaning techniques are generally applied, known as off-line (crank wash) and on-line cleaning. Under extreme fouling conditions, hand washing of the IGVs may have to be conducted if time permits. During overhauls, hand cleaning of the full axial compressor is most effective.

Off-line washing is almost always carried out with the aid of a detergent, and extremely effective power recovery can usually be achieved. However, it is important that the manufacturer's recommendations are followed with respect to water quality, detergent/water ratio, and other operating procedures. Typically, wheel space temperatures must be below 93°C (200°F) to avoid thermal shock, and the off-line water wash is done with the machine on crank. The downtime for a crank wash depends mainly on the time it takes for cooling the engine. Larger heavy-duty engines can take 8–10 h to cool whereas on light aeroderivative engines only 1.5–3 h may be needed because of the low metal mass. Off-line cleaning is most effective when carried out in several steps that involve the application of a soap and water solution, followed by several rinse cycles using water alone. Each rinse cycle involves the acceleration of the machine to ~50% of the starting speed, after which the machine is allowed to coast to a stop. A soaking period between each wash and rinse cycle is also very important, and allows the soapy cleaning fluid to penetrate into the fouling deposits, thus dissolving salts and emulsifying oil and grease components. A useful method of determining the effectiveness of the off-line wash, and perhaps the need for additional wash or rinse cycles, is to collect samples of the effluent water from all available drain ports. The samples can be checked visually for color and clarity (which should improve as dirt is removed), or by a simple conductivity meter to monitor the removal of salts.

On-line washing is now very popular as a means to control fouling by avoiding the problem from developing. The primary objective of on-line washing is to extend the operating period between off-line washes by minimizing the buildup of deposits in the compressor, and thereby reducing the on-going incremental power losses. On-line washing is performed with the unit in full operation, and techniques and wash systems have now evolved to a point where this can be done effectively and safely. Outages or shutdown periods are not required. Depending on the nature of the fouling material, on-line washing is sometimes performed with water alone. In most cases, however, the use of an approved cleaner (detergent) will improve the effectiveness of the washing operation. This is particularly true if the fouling material contains any quantity of oil or grease. Demineralized water quality is almost always specified by the OEMs for on-line washing, to avoid the possibility of introducing harmful trace metal contaminants (such as Na + K) into the combustion turbine.

Optimal compressor cleaning can normally be achieved by adopting a combined program of regular and routine on-line washing (for example every few days or weekly), plus periodic off-line washing during planned outages.

Two main types of cleaning agents (detergents) are available for compressor washing and are normally classified as "water-based" or "solvent-based" products. Most of the new-generation products contain surfactants, wetting agents, and emulsifiers, and involve either an aqueous or a petroleum-based solvent system. Both type products are normally supplied as concentrates, and are diluted on-site with water (typically 1 part cleaner with 4 parts water) to produce the cleaning fluid. Solvent-based cleaners have traditionally been recognized as being more effective in removing oil and grease deposits, but certain new-generation water-based cleaners are formulated to be equally effective. Most water-based products also have the advantage of being biodegradable, which is an increasingly important requirement within this industry.

A comparative summary between on-line and off-line cleaning is provided in Table A.7.

Off-Line (Crank) Washing

The basic objectives of off-line cleaning are to clean a dirty compressor and to restore power and efficiency to virtually "new and clean" values. When performed correctly, and provided the operating period between off-line washing is not too long (site specific), this type of cleaning will typically restore virtually 100% of the lost power and efficiency attributed to compressor fouling. However, irrespective of the compressor performance degradation actually encountered, experience has indicated that users of both base load and peaking gas turbines should incorporate a minimum of three or four off-line compressor cleanings per year in order to remove the salt laden deposits on the downstream stages.

Table A.7 Comparison Between Off-Line and On-Line Washing

Off-Line Washing	On-Line Washing
▪ *Objective*: To clean a dirty compressor ▪ Reaches all compressor stages ▪ Virtually full power recovery, approaches "new & clean" values ▪ Involves cool-down & shutdown period (12 to 36 h) ▪ Lost revenue during shutdown ▪ Optimum time for cleaning may not be convenient, especially with base-load plants ▪ Effluent water for disposal	▪ *Objective*: To Keep a clean compressor cleaner for longer ▪ Extends operating period between off-line cleaning, thus enhancing production ▪ About 1% power can be recovered per wash, with a frequent on-line cleaning program ▪ Primarily cleans IGVs (No effect after water evaporates) ▪ No shutdown/no lost revenue ▪ Optimum wash frequency is site specific ▪ No effluent water for disposal ▪ Maintains safe margin to surge line ▪ Reduces risk of blade corrosion

Off-line wet cleaning (also known as crank washing) is a typical "soak and rinse" procedure for which the gas turbine must be shut down and cooled. The compressor is rotated at crank speed while a cleaning fluid is injected via nozzles or jet lances. Hand-held jet lances were widely used in the past and are still fairly popular with some operators. However, permanently mounted off-line nozzle systems installed in the air intake plenum are now preferred, and are generally offered as standard by most of the major turbine manufacturers. Nozzle design, system operating pressure, and total mass flow parameters vary widely, however, between the different manufacturers.

The injected cleaning fluid is normally a mixture of chemical detergent and water. Both solvent-based detergents and water-based products are used, depending mainly on the type of fouling material found in the compressor and local plant experience.

After a soaking period the compressor is rinsed with a quantity of fresh water. The amount of rinse water required and the number of rinse cycles vary from site to site, according to the gas turbine model and the amount of dirt removed during the off-line wash. Note that demineralized water is usually not specified for off-line cleaning and fresh water quality is normally acceptable. Effluent water drained from the compressor has to be disposed of according to local regulations.

Typical water quality requirements for an off-line wash are

- Total solids (dissolved & undissolved) $\leq$100 ppm
- Total alkali metals (Na, K) $\leq$ 25 ppm
- Other metals which may promote hot corrosion (V, Pb) $\leq$ 1 ppm
- pH 6.5–7.5

This water would be used for cleaner dilution and also for rinsing.

Off-line crank washing systems should be designed to achieve the highest washing efficiency with the smallest injection mass flow. This is important for the following reasons:

- Gas turbine users are interested in minimizing the quantity of effluent water to be disposed of.
- Some users claim that off-line water effluent is transported up to the exhaust during the wash procedure, and may wet and soak into the expansion joint fabric—resulting in damage of the expansion joint by lowering its insulation properties.
- A lower off-line injection mass flow will also reduce the potential risk of trace metal contamination in exhaust systems, where selective catalytic reactors (SCRs) for NO_x reduction or CO catalysts are installed.
- A smaller off-line injection mass flow will significantly reduce the required size, volume, and cost of washing skids, as well as the overall water and cleaner consumption.

Important Considerations During Off-Line Washing

Wetting of IGVs: Effective wetting of the IGV's suction area can be achieved by using full cone jet spray nozzles. The number of nozzles will be defined by the area to be wetted, which is usually the area between two intake struts. The necessary off-line injection mass flow characteristic will therefore be determined by the area to be wetted and impacted by the jet spray, and the distance between nozzles and the IGVs. The injection pressure is generally between 5.5 and 6 bar (80–87 psi). As spray jet trajectory is subject to gravity, the nozzle should be designed to provide an adjustment of up to five degrees to compensate for the gravity effect.

Crank washing (soaking) and rinsing can be considered as a mechanical "erosion" of the deposit layer, and soaking time will allow the cleaner to penetrate and soften the deposit layers. Systems with high atomization pressure will have no impact pressure on the IGVs, because the spray pressure will have decayed ~20 cm from the nozzle outlet, and the atomized droplets will need to be carried by the relatively low air flow rate produced at crank

On line cleaning

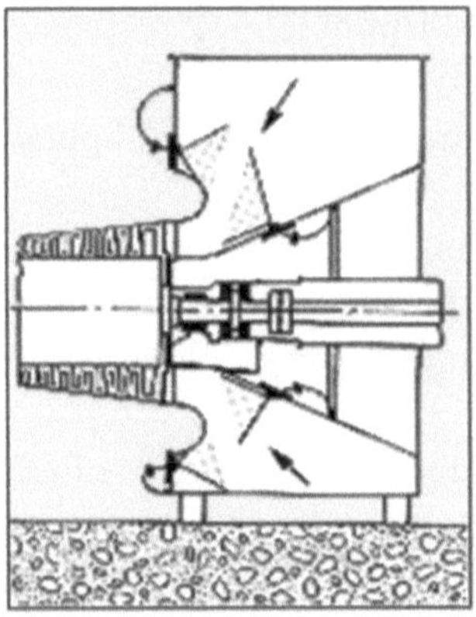

Typical arrangement with nozzles installed in two separate ring manifolds: spraying upstream and downstream.

Off line cleaning

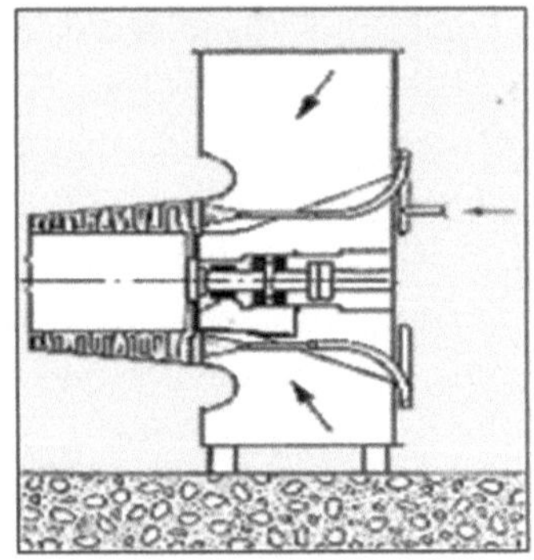

Typical arrangement with injection nozzles located upstream. This is a "soak and rinse" procedure. Also known as "Crank washing".

Fig. A.39 Typical location of wash nozzles.

speed. Most high-pressure systems do not show the same effectiveness in removing salts and insoluble compounds on downstream stages.

Typical location of wash nozzles for both on-line and off-line systems is shown in Fig. A.39.

Drainage: The effluent water collection system to drain and remove dirty water out of the engine is of prime importance. Equally important is the isolation scheme to prevent run-off water from penetrating sensitive areas such as sealing and cooling air systems, and instrumentation air systems. The physical location of bleed air taps on compressor casing is also important. If they are located on the bottom they are likely to drain run-off water, and therefore should preferably be located in the upper part of the casing. Drains in air inlet and compressor casing, combustion chamber, and exhaust should be located at the lowest point. The drain diameters should be sized for an easy run-off. Care must also be taken that they do not become plugged with dirt.

Engine speed variation: Off-line crank cleaning is very efficient for removing all deposits on all the compressor stages. To enable penetration of the wash and rinse fluid through the entire axial flow compressor, off-line washing should be conducted at *variable speeds*; for example by injecting the solutions during coast down of the shaft, after an acceleration of up to 500–600 rpm. By doing this, the pattern of the centrifugal forces on the injected solution through the compressor will decrease and allow better wetting and distribution on the blade and vane surfaces of all stages. By contrast, off-line washing at high and constant cranking speed will result in a lower cleaning efficiency. Conductivity measurements and checks on the clarity/turbidity of the drain water will help assess cleaning efficiency.

Procedures and precautions: OEM recommendations and checklists should be followed prior to a crank wash procedure. Some typical items of importance include

- Ensure that OEM's wheel space temperature criteria are met prior to the crank wash.
- Seal and atomizing air pipes should be closed-off to avoid water entry.
- IGVs should be in the open position (maximum airflow) prior to the crank wash.
- All drains should be opened.
- Flame detector valves should be closed.
- Auxiliary air compressor should be disconnected.
- Special precautions may have to be taken for regenerators, if these exist.
- If possible, the plenum should be hand washed to avoid additional dirt being washed into the compressor.
- Follow the OEM's or wash system supplier's recommendations regarding the duration and amount of fluid injection. These guidelines may need to be modified by evaluating the results.
- Approved anti-icing agents will be required if off-line cleaning is performed at ambient temperatures below or near to freezing. Even at crank speed, some degree of temperature depression will occur at the bellmouth.

Importance of the Rinse Cycle During Off-Line Washing

Efficient rinsing is extremely important to ensure that salts and other foulants dissolved or dislodged from compressor blading during the off-line wash cycle are thoroughly flushed out of the compressor casing. Ineffective or insufficient rinsing can cause foulants to be redeposited on the blading when the engine is restarted. Several rinse cycles may be required, and the best way to track the rinsing progress is to monitor the appearance (color) and conductivity of the effluent water, i.e., by taking effluent water samples from the compressor drainage system.

Effluent water samples taken during a typical off-line wash are shown in Fig. A.40, and it can be seen that conductivity is still very high after the first rinse, i.e., indicating that a high salt concentration remains within the compressor. However, after the second rinse cycle the effluent water conductivity is significantly reduced.

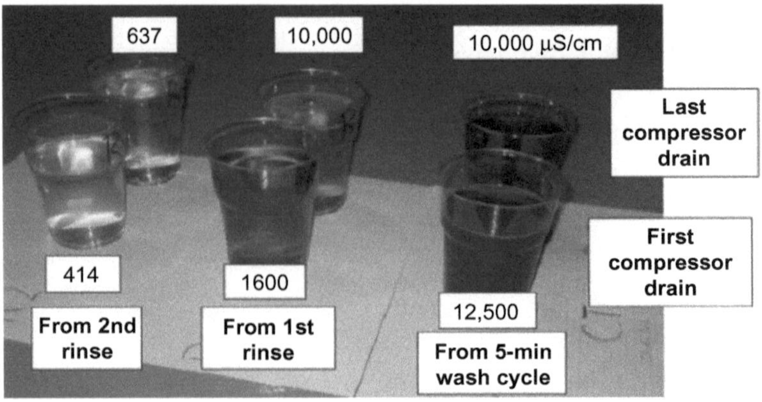

Fig. A.40 Effluent water samples taken during a typical off-line wash.

On-Line Washing

The basic objectives of on-line cleaning are to maintain the cleanliness of a compressor after off-line washing, to maintain power and efficiency by minimizing ongoing losses, and to extend the operating period between shutdowns required for off-line washing. On-line washing for fouling control has become increasingly important with base load combined cycle plants and CHP plants. It is also important for gas turbines in mechanical drive service, where little or no redundancy is installed and where the downtime associated with crank washing must be minimized.

On-line wet cleaning is performed while the gas turbine is in operation and at load. The procedure involves the injection of a mixture of water and chemical detergent via atomizing spray nozzles positioned around the compressor air intake plenum. This is followed by a flushing period using pure water. With on-line cleaning it is mandatory to use demineralized water for preparing the cleaning fluid and for flushing. This is because the turbine is in operation, and high temperature corrosion damage may occur if sodium or other contaminant metals enter the combustion path.

The water specification of the particular OEM should be followed, but typical values are presented below:

- Total solids (dissolved and undissolved) ≤ 5 ppm
- Total alkali and other metals which may promote hot corrosion (Na, K, Pb, V) ≤ 0.5 ppm
- pH 6.5–7.5

On-line washing should be performed frequently, for example daily or every 2–4 days depending on the severity of fouling at the particular site. Frequent on-line washing prevents the buildup and accumulation of deposits on IGVs and reduces the ongoing power degradation rate.

If the time interval between on-line washes is too long (for example greater than about 7 days), the benefits of this type of cleaning may not be seen. Furthermore, if fouling material is allowed to build up because of infrequent washing, there is a risk that salts and other contaminants might be released from upstream blading and pass through the engine as high concentration "slugs" when the next on-line wash is performed. It is for this reason that on-line washing programs should always be started on a clean engine, i.e., after an overhaul or crank wash.

Depending on the type of deposits (i.e., portion of water-insoluble compounds such as oil and grease), detergents may be used for every on-line wash, or for every second or third on-line wash, but not less frequently than once per week. Note that the longer detergent washing is not done, the greater the risk of downstream contamination due to large portions of insoluble compounds suddenly being removed when the next detergent wash is performed.

The duration of each on-line wash can also be varied according to the degree of fouling, engine size, plant experiences, etc. Typical on-line cleaning cycles are in the order of 10–20 min, and a flushing or rinsing cycle (using only demineralized water) of about the same duration should be applied after each cleaning cycle with detergent; for example 10–20 min cleaning cycle with detergent, followed by 10–20 min flushing cycle without detergent. This type of regular on-line wash regime will extend the operating period between outages required for off-line cleaning, which is particularly important for base load plants.

An important field study confirming the benefits of frequent on-line washing is reported by Schneider et al. (2009). The study was based on the monitored performance of six heavy-duty gas turbines (3 × 289 MW units and 3 × 180 MW units) operating at two different sites, and with different on-line wash regimes. The effectiveness of a daily on-line wash was found to be about twice as high as a weekly on-line wash. It was also found that with a daily on-line wash regime, detergent was required only once per week, and demineralized water alone could be used on the other days.

On-Line Compressor Cleaning Efficiency

Fouling of the first-stage guide vanes is the primary cause of reduced air mass flow through the compressor. On-line cleaning is most effective in removing such deposits, and therefore restoring design air mass flow and lost power. Regular on-line cleaning will keep inlet guide vanes clean and free from deposit buildup. Droplets of cleaning solution or water may survive up to the 6th stage, but most will have vaporized by then. On-line cleaning has no effect on downstream stages.

Fig. A.41 shows a compressor that was washed on-line every 4 days with detergent during 1 month of continuous operation. The first-stage guide vanes of the same compressor are shown in Fig. A.42. Note that this

Fig. A.41 Axial compressor washed on-line every 4 days during 1-month continuous operation.

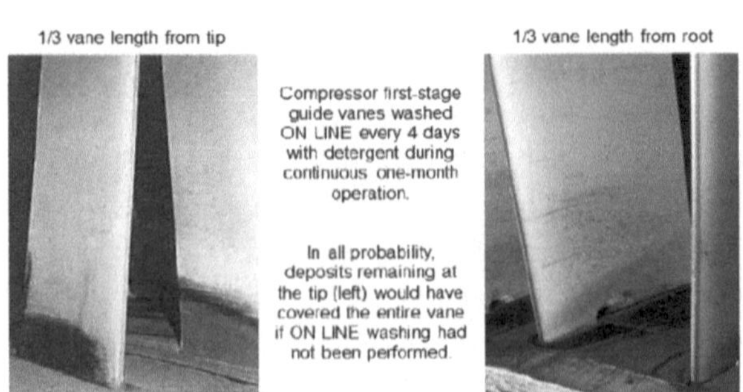

Fig. A.42 First-stage guide vanes washed on-line every 4 days during 1-month continuous operation.

photograph shows fairly heavy, black deposits still remaining at the tip and root areas of the IGVs, and this illustrates the severity of fouling that would have occurred over the entire vane length if regular on-line cleaning had not been performed.

Compressor Cleaning With Demineralized Water

As mentioned previously, compressor-fouling behavior is highly plant specific, and it can be very misleading to generalize about the efficacy of a particular wash program. For example, the use of demineralized water without detergent may prove to be an effective way to clean the compressor in some specific locations, but it would be wrong to state that this is a recommended method for all gas turbine sites. Using water as the on-line cleaning media will be effective if fouling deposits are totally water-soluble and/or water-wettable. However, if fouling deposits contain some quantity of oil or grease, water alone will not be effective. Foulants containing even trace quantities of hydrocarbon need the surfactant action of a detergent to break surface tension forces and lift the deposit from blade surfaces.

As a practical matter, and at the risk of "generalizing," we believe that there are few engines in the overall population that will have totally water-soluble or water-wettable deposits, and most operating environments are likely to involve some quantity of oil and grease. Using water exclusively as the cleaning media will typically result in very short time intervals between washings, in order to avoid any deposit build-up at all. Also, the use of demineralized water alone can be detrimental in cases where deposits are water insoluble, or where deposits are a combination of water-soluble and insoluble compounds. This is because the insoluble material will not be removed, which allows a greater buildup of deposits on the front stages.

Blade Wetting Considerations With On-Line Washing

Effective wetting of the IGVs is obtained by a uniform and finely distributed atomized cleaning solution. Droplets must be stable in size and small enough that they do not cause blade erosion (due to the high blade speeds during operation).

Droplets are subject to gravity, so they must also be light enough that they do not drop out of the air stream before they reach the compressor blade surface. Nonuniform wetting of the IGVs will result in spot cleaning and heavier droplets will most likely fall to the bottom, wasting some of the injected cleaning solution.

Environmental Impact of On-Line Washing

It is recognized that on-line washing will create a small increase in CO emissions due to disturbance of the combustion conditions associated with the injection of water. However, this should not be misinterpreted as a reason not to perform on-line cleaning. Any emission increase will be short-lived (only for the duration of the wash and rinse cycle), and is normally classified as a "transient condition" that can be addressed in the operating air permit in a similar way to startups and shutdowns. Low mass flow on-line injection systems will have less impact on CO emissions than high mass flow designs, because less water is injected. Also, the use of a detergent during on-line washing can be expected to have an insignificant impact on emissions, compared to the effect of water alone and the quantity of hydrocarbon fuel consumed.

Design Considerations for On-Line Wash Systems

The design of an on-line washing system should attempt to obtain the highest possible cleaning efficiency with the lowest injection mass flow rate, and this ***can*** be achieved in combination with an optimum washing regime as discussed above. Frequent on-line washing (to keep a clean compressor clean) enables the use of low injection mass flow rates, and this can only be considered as "good" for the gas turbine. To summarize:

- Frequent on-line cleaning with a low injection mass flow system minimizes the risk of deposits being suddenly washed from the front stages onto downstream compressor stages. This also addresses concerns that removed dirt may enter airfoil film cooling systems of turbine blades.
- Low injection mass flow reduces blade loading, and creates less stress on downstream blading.
- High water mass flow may interfere with flame detector intensities on units with Dry Low NO_x (DLN) combustors, and may also create a trip by causing fogging of flame detector lenses, etc.
- Higher CO emission levels may be observed during on-line washing, which becomes more significant with higher water injection rates.
- Typically, high mass flow nozzles create larger sized droplets. This means they are more influenced by gravity, and have a greater tendency to fall and be deflected onto inlet plenum surfaces. This also increases the risk of blade erosion.
- Effective on-line cleaning requires that the IGVs are thoroughly wetted with appropriately sized water droplets (typically 50–250 μm range). Any excess water is likely to "stream" over inlet plenum surfaces and struts,

etc., and serves little or no purpose in the washing process. As mentioned above, excess water impacting the root area of IGVs may also initiate erosion of the rotating blades.

- Lower injection mass flow rates reduce demineralized water and cleaner consumption, and also reduce the size, volume and cost of the wash skids.
- Experience has shown that a low water injection rate does not impair on-line cleaning efficiency.

Nozzle Location for On-Line Washing

Correct positioning and location of on-line injection nozzles are of prime importance to achieve uniform wetting and efficient blade cleaning. Nozzles should be designed to inject a small quantity of finely atomized cleaning solution into the air stream where it will be thoroughly mixed and carried uniformly into the compressor bellmouth. Compared to off-line cleaning these design factors are also extremely important in ensuring operational safety, because insufficient atomization and/or nonuniform injection flow can result in blade erosion and vibration. A relatively high number of on-line nozzles positioned in the air inlet casing both up-stream and down-stream of the bellmouth ensures a better distribution of the injected fluid into the air stream, and provides better wetting.

The design and configuration of air inlet systems vary according to different gas turbine models, and this is also an important consideration in selecting appropriate nozzle locations. For example, some inlet systems direct the entire incoming air stream from the top of the plenum, down toward the bellmouth. This creates a higher air velocity within the plenum, and correct positioning of the on-line nozzles is more challenging. Other inlet systems split the air stream so that it enters the plenum equally from the right- and left-hand sides. This design lowers the air velocity and creates fewer disturbances to the airflow pattern. Hence, positioning of the on-line nozzles is somewhat easier. A typical on-line nozzle manifold is shown in Fig. A.43.

Nozzle Designs for On-Line Washing

The on-line injection nozzle is the most critical component in the washing system, and its design determines the overall effectiveness of the cleaning process. Important issues such as nozzle location, injected mass flow, wash frequency, and the use of detergent can be rendered immaterial if the installed nozzles do not perform. A nozzle is essentially an atomizing device,

Fig. A.43 Photograph of on-line nozzle manifold. *(Courtesy of Turbotect Ltd.)*

and they are designed to create droplets in a certain size range and with a specific spray profile. However, in order to achieve the design droplet size and spray pattern, the nozzle must be operated at its design flow rate and operating pressure. Changing the design parameters of a particular nozzle will affect its spray characteristics. For example, droplet size will usually increase by reducing operating pressure below the design point.

Many different types and designs of on-line injection nozzle are used within this industry; some are available directly from the gas turbine OEMs and others are manufactured and marketed by independent suppliers. Design differences between the various types of nozzles are quite significant, and this results in a wide range of mass flow rates, operating pressures, spray patterns, droplet size ranges and, not surprisingly, performance characteristics. However, accepting that different design philosophies do exist (and probably will continue to exist), there is fairly common agreement within the industry that on-line droplet size should be within the range 50–250 μm. If droplets are too small, they are easily deflected by the air stream, and may not survive to perform wetting of the IGVs. If droplets are too large they are more influenced by gravity, and tend to fall toward the lower surfaces of the inlet plenum. Large-sized droplets are also more likely to cause erosion problems.

General design factors for on-line wash nozzle systems are as follows:

- Nozzles must provide excellent atomization characteristics, and create a stable distribution of droplets within the recommended size range 50–250 μm.

- Nozzles should be positioned to enable uniform wetting of the IGVs. (Note that high mass flow designs can create too much spray that collects and "streams" on plenum surfaces.)
- A sufficient number of nozzles must be installed to provide the required total mass flow for the particular gas turbine model.
- Flush-mounted, low profile nozzle designs create less disturbance to the air flow.
- All components must be corrosion resistant (stainless steel) and designed for maximum safety.
- The system should be easy to install on new units and on retrofits.
- The system should be operated at its design flow rate and pressure.

Fig. A.44 shows one particular design of on-line nozzle that is adjustable, and can be rotated in two dimensions to optimize the spray direction. This nozzle is a low-pressure, low mass flow design, and produces a 90° cone-shaped spray pattern. The nozzle's low profile ensures minimal penetration into the air stream to avoid airflow disturbance and vibration. It also prevents misuse of the nozzles as climbing supports during compressor intake inspections.

A useful review of gas turbine on-line wash systems has been published by Mund and Pilidis (2004). The different types of nozzles and on-line wash system designs used in this industry are surveyed, and categorized according to operating pressure and mass flow rate.

Fig. A.44 Adjustable on-line injection nozzle, for flush-mounting in the air inlet plenum. *(Courtesy of Turbotect Ltd.)*

Special Considerations for Large Gas Turbines

As gas turbines have become larger in output, and physical dimensions have grown considerably, the ability to effectively distribute droplets across a large inlet plenum and uniformly wet the blading has become a very important consideration. Due to the longer distances involved, gravity forces on the droplets become more significant and may create a deviation in the desired trajectory. Details are provided in Jeffs (1992, 2003) and Chellini (2004).

A new type of on-line nozzle that minimizes gravity effects and air mass flow distortion is shown in Fig. A.45. The nozzle produces a flat profile water spray that is "sandwiched" between two high velocity flat-profile air sprays. These dual air sprays protect the water spray trajectory, enabling it to break through the boundary layer into the main air stream and thereby preventing premature deflection of the droplets. The nozzle is a low pressure, low mass flow design. It operates at 4 bar pressure (58 psi) and produces droplets in the size range 50–250 μm. A nozzle in operation at base load on a Frame 9E gas turbine is shown in Fig. A.46.

In order to observe and verify the behavior of wash trajectories, it is a good idea to install a viewing window in large gas turbine ducts. One such window and associated lighting arrangement is shown in Fig. A.47.

Fig. A.45 Online wash nozzle with air curtain developed for large-sized gas turbines. *(Courtesy of Turbotect Ltd.)*

Fig. A.46 On-line wash nozzle in operation on a 123 MW gas turbine at base load. *(Courtesy of Turbotect Ltd.)*

Fig. A.47 Illuminated viewing window on a 123 MW gas turbine.

Combined On-Line and Off-Line Washing Approaches—Field Test Findings

An optimum-washing regime would combine on-line and off-line cleaning to minimize power deterioration and extend the operating period between outages required for crank washing. Depending on local site conditions, the number of off-line washes may be reduced to two or three per year, compared to four or more per year without on-line cleaning. Information on some important field studies involving combined on-line and off-line wash programs is presented below.

Combination of On-Line and Off-Line Washing—Tests on 2×30 MW Gas Turbines

While much of the information on the best approach to on and off-line washing is anecdotal, the first detailed and scientific investigation covering a long operating period was conducted by Stalder (1994, 1998). This study ran for over 4000 operating hours in the mid-1980s at the Energieproduktiebedrijf Utrecht (UNA) PEGUS 100 MW combined cycle plant in the Netherlands (2 × 30 MW gas turbines and 1 × 40 MW steam turbine). Salient environmental and operational conditions included

- The plant is situated 60 km from the sea
- A very busy motorway crosses over the canal near the plant
- Local industries include chemicals and food processing. Air pollutants include dust, salts, and fine aerosols
- The gas turbines ran on natural gas (no fouling of the hot section)
- All power measurements were made with the gas turbines running in temperature control mode at base load

Results showing power output changes over time (corrected to new and clean guaranteed conditions) are shown in Fig. A.48.

Key findings of this study include

- Without cleaning, the power output degradation tended to stabilize itself with increasing operating hours. It was confirmed in the unit tested that the power degradation stabilized at 90% base load (ref. new and clean).
- Power recovery after an off-line crank wash (soak and rinse wash) was significantly higher than after an on- line wash.
- On-line washes were performed at time intervals in the range between 700, 350, and 120 operating hours. Results showed that plant performance is significantly improved with shorter on-line washing intervals as these prevented incremental power degradation.
- The combination of both crank and on-line washing methods was found to be the most effective and economical.

Improvements With Shorter On-Line Washing Intervals; 6-Month Field Study on a 66 MW Gas Turbine

In the Spring of 1990, UNA and Turbotect jointly conducted a second long-term field test on a 66 MW natural gas fired turbine operating in the

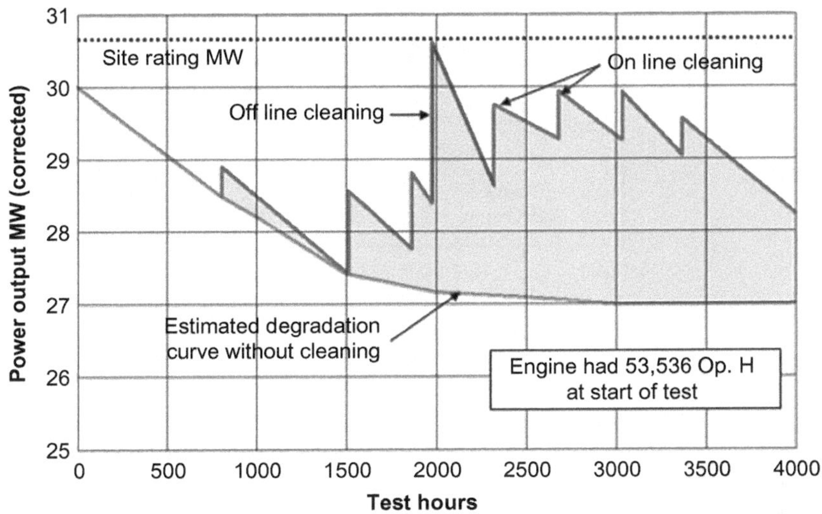

Fig. A.48 4000 Hour operating test with on-line and off-line cleaning on a 30 MW gas turbine (Stalder, 1998).

Lage Weide 5 combined cycle plant located at the same site. The tests were conducted over 18 months under the combined on-line and off-line wet cleaning regime, from 18 May 1990 to 18 November 1991. During the entire test period, the gas turbine unit operated for a total of 8089 h. An outage for a major overhaul (at 26,408 operating h) took place after 3915 operating hours since beginning of the test. The tests aimed to give some comparative indications of the effectiveness of more frequent on-line washing as applied to a new machine, and to one that operated for several years. The air inlet filtration system consisted of a weather louver, a first-stage coarse filter and a second-stage fine filter. Details are reported in Stalder and van Oosten (1994) with salient points highlighted below:

- On-line compressor washes were performed on average every 4 days at base load, with the gas turbine on temperature control mode. Gas turbine performance was measured before and after each wash. The pattern of the corrected power output is shown in Fig. A.49.
- Observations showed a sustained high output level close to the nominal guaranteed rating, despite severe atmospheric conditions.
- In the first evaluation block of Fig. A.49, the gas turbine was operated at a load factor of 97.6% or 2.4% below the original guaranteed site power output at new and clean conditions. During this period 38 on-line compressor washes were performed, at an average frequency of one every

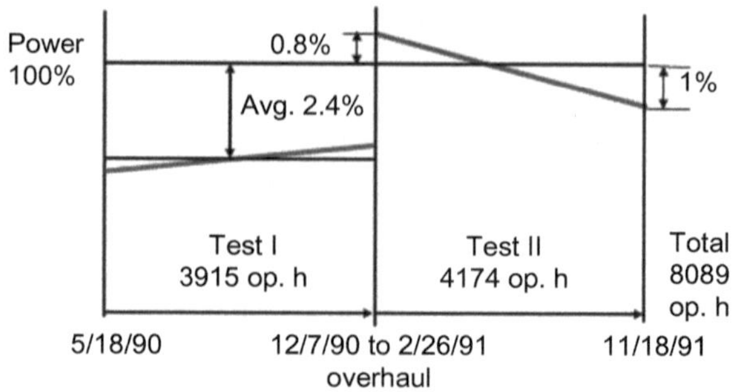

Fig. A.49 Summary of corrected power output during 8000 h test with on-line and off-line cleaning on a 66 MW gas turbine.

4 days. In addition, three off-line washes were performed by taking the opportunity whenever the gas turbine plant was shut down for a few days. This was after intervals of 760, 2435, and 605 operating hours. The average increase in power output after an off-line wash was ~1800 kW. The trend (power versus time) of the performance tests made during this period is nearly horizontal, showing that aging due to mechanical wear and tear of the gas turbine (nonrecoverable deterioration) had already stabilized.

- In the second evaluation block of Fig. A.49 the gas turbine started in a practically new and clean condition as the result of work done during the major overhaul. The corrected results of the compressor wet cleaning regime in the second evaluation block show that the gas turbine operated for 4174 h at a load factor of 100.16% or 0.16% above the original guaranteed site power output at new and clean conditions. At the end of this period the number of operating hours of the gas turbine was 30,725. During this second period, 45 compressor on-line washes were performed, again, at an average of one every 4 days. In addition, two off-line washes were performed, one after 1143 and the second 1381 operating hours later. The average power output increase after each offline wash in the second period was ~1 MW.

Findings of the Study

The following findings were reported by Stalder (1998):

- Out of the 83 on-line washes made during the total testing period covering 8089 operating hours, 87% (i.e., 72 on-line washes) demonstrated positive power recovery.

- The average power output recovery measured after an on-line wash was 712 kW. This relative small amount represents ~1% of the nominal gas turbine power output.
- No definitive results on efficiency improvements could be demonstrated because of incomplete data over the testing period with regard to fuel gas analysis and densities to determine the lower heating value, which is required for accurate heat balance calculations to derive corrected efficiencies and turbine inlet temperature.
- Power recovery due to off-line cleaning if the unit's performance is close to nominal guaranteed values, is not as significant as an on-line wash under deteriorated conditions.
- On-line and off-line cleaning is complementary.
- Shutdowns and start-ups can positively affect compressor fouling by causing deposits to spall off. The deposits may soak up humidity during a standstill and the swollen material will partly spall-off as the compressor is accelerated during gas turbine start up.
- The test program confirmed that frequent on-line cleaning extended the time interval between off-line cleaning operations. This is a real benefit to the operator, because the scheduled down time allowed for maintenance can be reduced if the frequency of off-line cleaning with its associated cooling down time can be reduced.
- The results demonstrated that a combined on-line and off-line washing regime can effectively be applied to new or old engines. The unit performance measured in the second evaluation period shows that the power output trend most probably followed the aging of the gas turbine.

In terms of power recovery and economic return, the optimum program is likely to be a combination of regular on-line and periodic off-line wet cleaning, coupled with the use of appropriately designed equipment systems and effective chemical detergents. Many base load power plants now adopt this combined cleaning regime and experience proves that the concept is highly successful in terms of cleaning efficiency and restored power.

To summarize, an optimum cleaning program is site- specific, and will be influenced by the severity of compressor fouling, the magnitude of associated power losses, and the operating schedule of the particular plant. Program parameters and options to be considered and evaluated include the following:

- Cleaning technique, i.e., on-line, off-line, or combined on-line/off-line wet cleaning. For most plants the combination technique appears to be optimal.

- Frequency of each on-line and off-line cleaning cycle
- Duration of each cleaning cycle and total mass flow of injected cleaning fluid
- Duration of the rinsing or flushing cycle and the amount of rinse water used
- Selection of the most appropriate water-based or solvent-based chemical detergent

The possibility of ice formation during cold weather washing operations should not be overlooked, and the use of an approved anti-icing agent can be considered.

Compressor Wash Skids

Wash skids come in a wide variety of designs, engineering concepts, and sizes. Some utilize pumps and others use compressed air. Skids can be designed for only off-line washing, or for combined off-line and on-line duty. It is important that the design of the wash skid be integrated with the nozzle systems installed on the gas turbine so that wash fluid and rinse water are delivered at the correct pressure, temperature, and flow rate. It is also important that the user make a design review of the skid from an operability and reliability standpoint.

Some important considerations are as follows:

- Skids should be designed to provide sufficient storage capacity of cleaning fluid (i.e., water–detergent mixture) and rinse water to enable at least one complete off-line or on-line washing operation on one gas turbine, without the need to refill the storage tanks. In most cases the volumes required for off-line cleaning will exceed those for on-line cleaning, so the crank wash volumes will normally determine tank sizing.
- Skids can be designed to serve a single gas turbine or multiple units on the same site. If multiple gas turbines are involved, outlet manifolds for connection to off-line and on-line piping to each gas turbine should be included. If possible, skids should also be located so that piping runs to each gas turbine are approximately equal.
- Skids can be designed with a variety of control options: For example "local manual" control for off-line washing, and fully automatic control (local and/or remote) for on-line washing. Note that for automatic on-line washing, a control system can be provided so that the wash sequence is initiated via a signal from the plant control room.

- The complexity of the wash skid control system can be a fairly major cost factor, and users should determine up front whether options for fully automatic operation are really required.
- Certain gas turbine models need an external compressed air source for opening and closing shut-off valves and drain valves, as required for off-line washing. In this case, wash skids can also be specified to include an air compressor system.
- Transfer of wash fluid and water to the nozzle systems can be facilitated by an electrically driven pump, or by compressed air. In the case of compressed air systems, the equipment must be designed or sized to enable complete wash and rinse cycles to be performed without interruption and without pressure fluctuations. It is also extremely important that compressed air used for washing systems is absolutely oil-free. It is not uncommon to find situations where oil is being introduced into the compressor via an air-driven wash system.
- Some OEMs recommend that off-line washing be performed with hot water, and in this case the skid must be designed with a heating system; which could be steam or electrical. This requirement greatly increases the physical size and cost of the wash skid. The purpose of heating the off-line wash fluid is to reduce gas turbine cool-down time, but it is debatable as to whether cleaning efficiency is actually improved.
- In cases where several gas turbines cannot be served by one centrally located stationary wash skid, and as a less expensive alternative to installing multiple skids, a mobile wash cart is often considered. Wash carts, however, have limited storage capacity for cleaning fluid, and usually have no storage tank for rinse water, so more filling and handling time is required. As a general rule, most operators consider wash carts to be inconvenient, and certainly more man-hours are required for compressor washing than with stationary skids.
- Other considerations that apply to both wash skids and mobile carts include materials of construction, safety features, reliability, pressure testing (especially for high pressure systems), hose couplings, interconnecting piping, maintenance, and operating parameters. The handling, storage, and mixing of chemical detergents (usually supplied as concentrates) also need to be considered.

A typical on-line/off-line wash skid designed to accommodate 5 × heavy-duty gas turbines is shown in Fig. A.50.

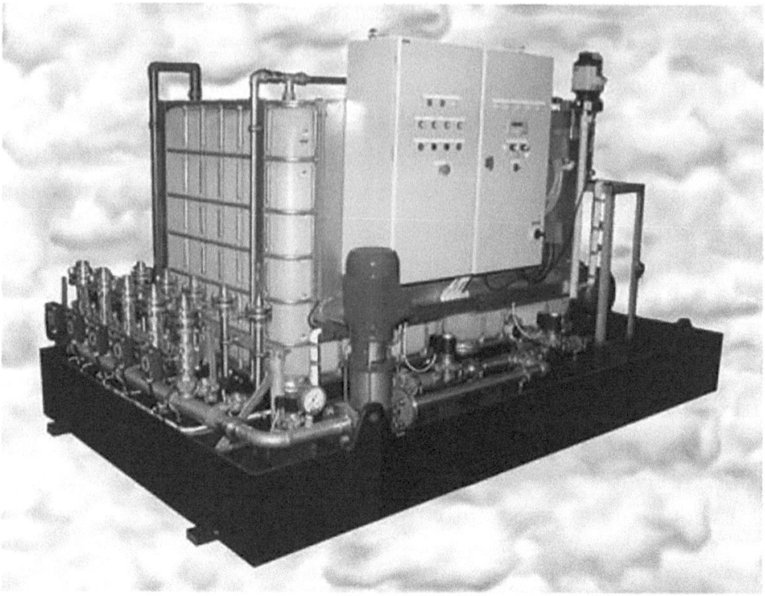

Fig. A.50 Typical on-line and off-line wash skid *(Courtesy of Turbotect Ltd.)*

Compressor Wash Fluids and Detergents

Detergents greatly enhance the removal rate of compressor deposits, compared to the use of water alone. Their function is primarily to reduce surface tension forces between deposits and blade surfaces (and between individual deposit particles), and also to emulsify oil and grease components. Actual removal mechanisms are likely to involve both dissolution and emulsification (depending on the nature of the deposit), coupled with the centrifugal forces and air mass flow effects within the compressor.

Most fouling deposits are mixtures of water-soluble, water-wettable, and water-insoluble materials. Water-soluble compounds can cause corrosion, since they are usually hygroscopic and will absorb moisture and acidic airborne pollutants such as SO_x and NO_x. In fact, pH values of 4 and lower can often be measured in compressor blade deposits. Soluble chloride salts are often found, and can promote pitting corrosion. They can be rinsed, but they can also be embedded within water-insoluble compounds. Water-insoluble compounds are mostly organic such as hydrocarbon residues, and these materials act as "glue" to entrap and hold other foulants. All compressor deposits become more difficult to remove if left untreated, as the aging process bonds them more firmly to the airfoil surface and thus reduces cleaning efficiency.

The benefits of using a detergent during off-line cleaning are well established throughout the industry, and almost all crank wash procedures specify the use of a cleaning agent during the wash cycles. However, the benefits of using a detergent for on-line cleaning are less well appreciated, and in fact many of the gas turbine OEMs send "mixed signals" to the industry when confronted with this question. It can also be quite misleading to generalize on these type questions, because the nature and degree of compressor fouling is extremely site specific, and a cleaning program that is effective in one location may not work in others. Water alone can indeed be effective in removing fouling material that is completely water-soluble and/or water-wettable, but it is unlikely to have any real effect on deposits containing oil and grease. For these situations, and probably in fact for the majority of cases, a chemical detergent is needed to counteract oil and grease components so that complete removal of all the fouling material can be ensured. In fact, we believe that there are few engines in the overall population that will have totally water-soluble or water-wettable deposits, and most operating environments are likely to introduce at least some quantity of oil and grease.

As discussed above, we also believe that using demineralized water alone can be detrimental in cases where deposits are a combination of water-soluble and water-insoluble compounds. This is because the insoluble material will not be removed, thus allowing a greater buildup of deposits on the front stages. Provided a detergent meets the strict quality specifications established by the OEM for on-line compressor cleaning, we believe there is no technical reason to limit or restrict its use.

There are many different brands of compressor cleaners available on the market, but generally they are classified as being either "*water-based*" or "*solvent-based*" products.

Traditionally, solvent-based cleaners have been recognized as being more powerful in terms of oil and grease removal, but their popularity within the industry has now declined because of environmental and safety considerations. During an off-line crank wash, for example, the discharged effluent water will usually require treatment before disposal and may be classified as hazardous waste. Also, due to the presence of hydrocarbon solvents, this type of cleaner may be subject to transportation restrictions and/or higher freight rates.

Water-based cleaners are now more widely applied within this industry, and are generally preferred because of their safe handling properties and high level ofbiodegradability. Many water-based products can be highly effective in removing a wide variety of foulants, and provide acceptable cleaning performance in most situations. New-generation water-based formulations are also available that can match the cleaning efficiency of solvent-based products.

Compressor cleaners must comply with various industry standards (for example MIL-specs and environmental regulations), as well as quality and use specifications established by the various gas turbine manufacturers. General requirements include

- Low trace metal impurity content (extremely important for on-line applications).
- Thermal decomposition properties that ensure gums or sticky residues do not form during the cleaning operation. (Again, this is particularly relevant for on-line cleaning applications.)
- Approximately neutral pH value.
- High cleaning performance, which can be evaluated via several established testing procedures.
- Long-term storage stability.
- Compatibility with alloys, coatings, paints, and elastomers within the gas turbine engine.

Many detergents are sold as concentrates, and should be used at an appropriate dilution ratio as recommended by the manufacturer. A typical dilution ratio is one part detergent to four parts water.

On-Line and Off-Line Compressor Cleaners

Cleaners available on the market today are generally nonionic, and designed to fulfill gas turbine engine manufacturer's specification for both on-line and off-line cleaning. This also simplifies stock keeping and on-site handling. The main constituent of a cleaner is its surfactant (surface active agent), the purpose of which is to reduce surface tension of the cleaning fluid to enable it to wet, penetrate, and disperse the deposits. The surfactant also reduces surface tension of oil and grease films, which otherwise have strong adherence to blade surfaces and to other fouling materials. Such rapid penetration and dispersion of the deposits cannot be achieved with water alone. Surfactants are therefore needed for water-insoluble deposits (both liquid and particulate types), to enable their removal from compressor blade surfaces and to prevent redeposition.

Foaming of Compressor Cleaners

Foam is an indication of the degree of activity of the compressor cleaner, and therefore of the effectiveness of the surfactants used in the cleaner formulation. Foam helps to achieve a better distribution and penetration of the cleaning solution into the deposits during off-line washing, and it extends contact time by holding moisture. Water films alone will tend to drain off the blades more rapidly, thus reducing contact time during the off-line soaking period. In addition, the foam also acts as a dirt carrier, to help transport the removed foulants from the compressor. Foam should be quickly displaced during rinsing, but if it dissipates (collapses) too rapidly it can release and redeposit dirt within the compressor. So-called low-foaming cleaners can reduce the time required for rinsing after an off-line wash, but insufficient foam or foam which collapses too rapidly can also impair the cleaning process.

Corrosion Inhibitors

The concept of including a corrosion inhibitor in compressor cleaner formulations originated primarily in the aero industry, and may be important for jet engines that are frequently not re-operated (i.e., dried) immediately after compressor crank washing. This philosophy spread to the industrial gas turbine market, and as a result, many compressor cleaners used for stationary gas turbines are also required to include a corrosion inhibitor in order to comply with certain industry specifications.

Conductivity or pH measurements on off-line effluent water have shown that corrosive salts are often left behind to some extent after the cleaning cycle with detergent, and in this regard the corrosion inhibitor can provide some degree of protection. However, rinsing with water alone will also be beneficial in removing residual salt solution, and it is important that sufficient rinse cycles are performed. We also recommend that a "blow out" and "dry out" run is conducted after completing the rinse cycles, even if the cleaning agent contains a corrosion inhibitor.

Corrosion inhibitors have a high affinity to surfaces and will tend to form a film on compressor blading. However, there is a potential risk during online cleaning that this film may decompose in the mid-stage temperature range of 200°C (about 390oF). We therefore recommend that a demineralized water rinse of the same duration as the cleaning cycle is performed to eliminate any salts left behind. As discussed in previous chapters, this does not pose a risk to the gas turbine provided on-line washing is not applied to heavily fouled units.

Use of Hot Water for Crank Washing

Some OEMs recommend the use of hot water for off-line washing, and primarily this is to reduce the cool-down time before crank washing can commence so that thermal shock is prevented. However, this means that compressor wash skids need to be equipped with a heating system, insulated tanks, insulated piping, etc., and this significantly increases acquisition costs. Equipment maintenance costs and operating costs are also increased, and off-line washing must be scheduled in advance to enable heat-up time.

In addition to reducing cool-down time, it is conceivable that hot water may improve cleaning efficiency by helping to soften deposits and increase solubility. It is arguable, however, as to whether or not this improvement is significant, and probably more frequent washing (combined on-line & off-line regime) and the use of a good detergent will more than compensate for these possible benefits.

Engdar et al. (2004) recently addressed this subject, and conducted CFD studies on hot water injection for off-line compressor washing. The authors concluded that, regardless of its inlet temperature, injected wash fluid is cooled down to ambient air temperature well before the spray reaches the inlet guide vanes. This study indicates that heating the wash fluid may serve little benefit.

Cold Weather Compressor Washing

The ability to perform compressor washing during cold weather periods has become increasingly important for many gas turbine operators and OEMs. Occasions when it may be required are as follows:

- Off-line cleaning prior to an important performance test, regardless of ambient conditions.
- Scheduled outage for off-line cleaning may occur during winter months and cannot be postponed.
- Plant decisions to continue an on-line cleaning program during winter months.
- Plant decisions to perform on-line cleaning in order to delay an outage for off-line cleaning.

In the case of on-line cleaning it is important to remember that ice formation can occur in the compressor at ambient temperatures significantly above the freeze point. This is due to a static temperature depression in the bellmouth area, resulting mainly from air acceleration effects. The amount of temperature depression is related to specific design parameters of the compressor and can be as high as 10°C (18°F). Thus, depending on the particular model of gas turbine, conditions at the IGVs can be 10°C (18°F) lower than actual outside air temperature. The influence of static temperature depression is less significant with off-line washing due to relatively low crank speeds, but a few degrees of safety margin should still be allowed.

Various chemicals have been used as anti-icing agents—such as alcohols, ketones, and glycols. However, not all are suitable for gas turbine applications and a summary of issues and selection criteria are provided in Table A.8. Antifreeze agents based on propylene glycol offer significant safety and handling advantages (nonhazardous, high flash point, biodegradable, etc.), and can be used for both on-line and off-line cleaning applications. In essence they are used as a substitute for water during the cleaning and rinsing cycles. Using this type of anti-icing agent, on-line cleaning can be performed in ambient temperature conditions down to about −10°C (14°F) and off-line cleaning can be conducted at ambient temperatures down to approximately −22°C (−8°F). Actual product freeze

Table A.8 Issues With Antiicing Agents for Low Temperature Compressor Washing

Antiicing Agent	Issues and Comments
Alcohols (methanol, ethanol, isopropanol)	• Main problems are safety related; i.e., low flash point, volatility, and evaporation risk • Explosion-proof equipment and area rating required
Ketones (acetone, methyl ethyl ketone)	• Even greater safety risks due to low flash point
Ethylene glycol	• Does not present a flash point problem, but forms sticky deposits on compressor blades
Propylene glycol	• Nonhazardous, high flash point material • Nontoxic and biodegradable • Does not form sticky deposits during on-line cleaning • Generally recognized as most suitable for GT applications • PG antifreeze agents are available with freeze points down to −18°F (−28°C) • Can be mixed with certain compr. Cleaners to provide the wash fluid, or used "as delivered" for compressor rinsing

points can vary, however, depending on the specific formulation, so manufacturers' product information should always be consulted. Cold weather washing procedures and minimum allowed ambient temperatures also vary according to the different gas turbine manufacturers and for different gas turbine models, so OEM recommendations should always be observed.

Conclusions

This paper provides a comprehensive overview of compressor fouling and washing for gas turbine engines and presents, in a single document, the state of the art regarding the causes, effects, and control of axial compressor fouling. The paper has covered numerous practical aspects of direct applicability to users.

With increasing fuel costs and a highly competitive market, the understanding, measurement, and control of fouling deterioration are an imperative. Fouling rates can vary from plant to plant and are highly environment and machine specific. Furthermore, fouling behavior is influenced by inlet air filter selection and maintenance. Site weather patterns also have a dramatic impact on fouling. A judicious combination of off-line and on-line cleaning usually provides the best results in helping operators fight this common and insidious operating problem. Close monitoring of compressor performance can help optimize compressor-washing regimes and improve plant profitability.

Cyrus Meher-Homji is an Engineering Fellow and Technology Manager at Bechtel Corporation assigned to the LNG Technology Center of Excellence as a turbomachinery advisor to ongoing LNG projects on the aeromechanical design, selection, and testing of large compressors and gas turbines. His 33 years of industrial experience cover gas turbine and compressor application, design, and troubleshooting. Cyrus is a Fellow of ASME, and is active in several committees of ASME's International Gas Turbine Institute. He has a Master's Degree in Engineering from Texas A&M University and an MBA from the University of Houston. Cyrus is a member of the Texas A&M University Turbomachinery Symposium Advisory Committee.

Andrew Bromley is Vice President of Operations for Turbotect (USA) Inc. where he is responsible for the company's US activities in the areas of gas turbine compressor cleaning and fuel treatment technology. Andrew has over 30 years of experience with gas turbines and has specialized in the areas of high temperature corrosion, fuel chemistry, fuel treatment, hot gas path and compressor fouling, compressor washing, and other chemical applications associated with gas turbine performance. Mr. Bromley holds a diploma in Chemistry and a Master of Science degree in Corrosion Science & Engineering from City of London Polytechnic, United Kingdom. He is a member of the Royal Society of Chemistry and the Institute of Metallurgy & Materials (United Kingdom), and a Fellow of ASME.

Jean-Pierre Stalder is Chief Engineer and R&D Manager with Turbotect Limited in Baden, Switzerland. His responsibilities include the conceptual development of compressor cleaning systems and chemicals to alleviate gas turbine power degradation and efficiency loss due to compressor fouling. His career in turbomachinery has spanned over 30 years and is underlined by a number of experimental studies to improve the understanding of axial compressor fouling phenomena and its control. Other areas of expertise are the development and application of lubricity improvement additives for gas turbines operated on low viscosity liquid fuels, the combustion of alternative fuels, and high temperature corrosion inhibition for engines operated on crude oil, HFO or other contaminated fuels. Mr. Stalder is a mechanical engineer and a Fellow of ASME.

References

Aker, G.F., Saravanamutto, H.I.H., 1988. Predicting gas turbine performance degradation due to compressor fouling using computer simulation techniques. In: ASME Gas Turbine and Aeroengine Congress, Amsterdam, June 5–9, 1988, ASME Paper 88-GT-206.

Arnulfi, G.L., Massardo, A.F., 1993. The effect of axial flow compressor deterioration on gas turbines in combined cycle power plants. In: ASME IGTI Volume 8, Cogen Turbo Proceedings, pp. 607–614.

Bagshaw, K.W., 1974. Maintaining cleanliness in axial flow compressors. In: 1st Gas Turbine Operations and Maintenance Symposium, National Research Council-Canada, October, 1974.

Bammert, K., Woelk, G.U., 1980. The influence of blading surface roughness on the aerodynamic behavior and characteristic of an axial compressor. Trans. ASME J. Eng. Power 102 (April), 283–287.

Bromley, A.F., 2012. Gas turbine power degradation and compressor washing, tutorial part 2. In: ASME International Gas Turbine and Aeroengine Congress, Copenhagen, Denmark, June 11–15, 2012, ASME Tutorial No. GT2012–70231.

Chellini, R., 2004. Wash Nozzles Designed for Larger Turbines. Diesel & Gas Turbine Worldwide May.

Cleaver, R.E., 1988. Retrofitting of gas turbine air filtration systems using improved technology. In: Seminar on Turbomachinery, Paper A-1, BHRA Fluid Engineering Center, Milton Keynes, UK, October 27–28.

Cleaver, R.E., 1990. Gas turbine air filtration in tropical environments. In: Turbomachinery Maintenance Congress, October.

Dundas, R.E., 1982. The use of performance monitoring to prevent compressor and turbine failures. In: ASME International Gas Turbine and Aeroengine Congress, Paper No: 82-GT-66.

Dundas, R.E., 1986. A study of the effect of deterioration on compressor surge margin in constant speed, single shaft gas turbines. In: 1986 ASME Gas Turbine Congress, ASME Paper No: 86-GT-177.
Dundas, R.E., Sullivan, D.A., Abegg, F., 1992. Performance monitoring of gas turbines for failure prevention. In: ASME International Gas Turbine and Aeroengine Congress, Cologne, Germany, June 1–4, 1992., ASME Paper No: 92-GT-267.
Dusatko, R.A., 1995. Monitoring gas turbine performance. Power Eng. October.
Engdar, U., Genrup, M., Orbay, R.C., Klingmann, J., 2004. Investigation of the two-phase flow field of the GTX100 compressor inlet during off-line washing. In: ASME Turboexpo 2004, Vienna, Austria, June 14–17, 2004, ASME Paper No: GT2004-53141.
Felix, P.C., Strittmatter, W., 1979. Analysis of air pollution on the erection site of a Brown Boveri gas turbine. Brown Boveri Rev. 66, 97–103. February.
Flashburg, L.S., Haub, G.L., 1992. Measurement of combustion turbine non recoverable deterioration. In: ASME International Gas Turbine and Aeroengine Congress, Cologne, Germany, June 1–4, 1992. ASME Paper Number 92-GT-264.
Gidley, D., Fritz, F.M., Yu, H., 1993. The selection process and comparative air filter performance testing for combustion turbine inlet air filters. In: ASME International Gas Turbine and Aeroengine Congress, Cincinnati, Ohio, ASME Paper No. 93-GT-219.
Goulding, C.H., Rasmussen, M.G., Fritz, F.M., 1990. Technical and other considerations for the selection of inlet air filtration systems for high-efficiency industrial gas turbines. In: International Gas Turbine and Aeroengine Congress, Brussels, Belgium, June 11–14, 1990. ASME Paper No: 90-GT-176.
Haskell, R.W., 1989. Gas turbine compressor operating environment and material evaluation. In: ASME International Gas Turbine and Aeroengine Congress, Toronto, Canada, June 4-8, 1989, ASME Paper No: 89-GT-42.
Hsu, L.L., 1988. Total corrosion control for industrial gas turbines: airborne contaminants and their impact on air/fuel/water management. In: ASME International Gas Turbine & Aeroengine Congress, Amsterdam, June 6-9, 1988. ASME Paper No. 88-GT-65.
Jeffs, E., 1992. Compressor washing for large gas turbines. Turbomach. Int. Mag. (Sept/Oct).
Jeffs, E., 2003. Turbotect's innovative on-line wash nozzle for large gas turbines. Turbomach. Int. Mag. (May/June).
Kimm, M.H.P., Langlands, D., 1985. Gas turbine intake filter systems related to offshore platform installations. In: ASME International Gas Turbine and Aeroengine Conference, March 18–21, 1985, Houston, Texas. ASME Paper No. 85-GT-109.
Koch, C.C., Smith, J.H., 1976. Loss sources and magnitudes in axial compressors. Trans. ASME J. Eng. Power 98 (July), 411–424.
Kurz, R., Brun, K., Wollie, M., 2008. Degradation effects on industrial gas turbines. In: Proceedings of ASME Turbo Expo 2008: Power for Land, Sea and Air, June 9-13, 2008, Berlin, Germany; ASME Paper No. GT2008-50020.
Kurz, R., Brun, K., 2000. Degradation in gas turbine systems. In: ASME International Gas Turbine and Aeroengine Congress, Munich, Germany, May 8-11, 2000, ASME Paper No: 2000-GT-345.
Kurz, R., Brun, K., 2012. Fouling mechanisms in axial compressors. Trans. ASME J. Eng. Gas Turbines Power. 134. GTP-11-1153.
Lakshminarasimha, A.N., Boyce, M.P., Meher-Homji, C.B., 1994. Modeling and analysis of gas turbine performance deterioration. Trans. ASME J. Eng. Gas Turbines Power 116 (January), 46–52.
Manfrida, G., et al., 1988. Experimental study of flow in gas turbine inlet duct models. Quaderni Pignone 44, 19–26.

Meher-Homji, C.B., Bromley, A.F., 2004. Gas turbine axial compressor fouling and washing. In: Proceedings of the 33rd Turbomachinery Symposium, September 2004, Houston, Texas.

Meher-Homji, C.B., Chaker, M., Bromley, A.F., 2009. The fouling of axial flow compressors—causes, effects, susceptibility and sensitivity. In: Proceedings of the ASME International Gas Turbine Congress, Turboexpo 2009, Orlando Florida, June 8-12, 2009, ASME Paper No: GT2009-59239.

Meher-Homji, C.B., 1990. Gas turbine axial compressor fouling - a unified treatment of its effects, detection and control. ASME Cogen Turbo IV, New Orleans, LA, August 27–29, 1990, Also inInt. J. Turbo Jet Engines 9 (4), 99–111. 1992.

Meher-Homji, C.B., Gabriles, G., 1998. Gas turbine blade failures—causes, avoidance, and troubleshooting. In: Proceedings of the 27th Turbomachinery Symposium, Houston Texas, September.

Meher-Homji, C.B., Chaker, M., Motiwalla, H., 2001. Gas turbine performance deterioration. In: Proceedings of the 30th Turbomachinery Symposium, Houston, Texas, September 17-20.

Meher-Homji, C.B., Lakshminarasimha, A.N., Mani, G., Boehler, P.D., Dohner, C.V., Ondryas, I., Lukas, H., Cincotta, G.A., 1993. Durability surveillance of advanced gas turbines—performance and mechanical baseline development for the GE frame 7F. In: International Gas Turbine and Aeroengine Congress, Cincinnati, Ohio, May 24-27, 1993, ASME Paper No: 93-GT-276.

Mund, F.C., Pilidis, P., 2004. A review of gas turbine washing systems. In: ASME Turboexpo 2004, Vienna, Austria, June 14–17, 2004, ASME Paper No: GT2004-53224.

Schaffler, A., 1980. Experimental and analytical investigation of the effect of Reynolds number and blade surface roughness on multistage axial flow compressors. ASME J. Eng. Power 102 (January), 5–11.

Schneider, E., Demircioglu, S., Franco, S., Therkorn, D., 2009. Analysis of compressor online washing to optimize gas turbine power plant performance. In: Proceedings of the ASME International Gas Turbine Congress, Turboexpo 2009, Orlando Florida, June 8-12, 2009, ASME Paper No: GT2009-59356.

Schweiger, T., 1983. North Sea operating experience-air intake design. In: Proceedings of the Fifth on Gas Turbines Operations and Maintenance, National Research Council of Canada, Calgary, September.

Scott, J.H., 1979. Axial compressor monitoring by measuring intake air depression. In: Proceedings of the Third Symposium on Gas Turbine Operations and Maintenance, National Research Council of Canada symposium, September.

Scott, J.H., 1986. Reduced turbomachinery operating costs with regular performance testing. In: ASME International Gas Turbine and Aeroengine Congress, June 1986, ASME Paper No. 86-GT-173.

Seddigh, F., Saravanamuttoo, H.I.H., 1990. A proposed method or assessing the susceptibility of axial compressors to fouling. In: ASME International Gas Turbine and Aeroengine Congress, June 11-14, 1990, ASME Paper No: 90-GT-348.

Stalder, J.P., 1998. Gas turbine compressor washing state of the art-field experiences. In: ASME International Gas Turbine and Aeroengine Congress, Stockholm Sweden, June 2-5, 1998, ASME Paper No: 1998-GT-420.

Stalder, J.P., Sire, J., 2001. Salt percolation through gas turbine air filtration systems and its contribution to total contaminant level. In: Proceedings of the Joint Power Generation Conference, June 4-7, 2001, New Orleans, LA., Paper JPGC2001/PWR-19148.

Stalder, J.P., van Oosten, P., 1994. Compressor washing maintains plant performance and reduces cost of energy production. In: ASME International Gas Turbine and Aeroengine Congress, The Hague, Netherlands, June 13-16, 1994, ASME Paper No: 1994-GT-436.

Stone, A., 1958. Effects of stage characteristics and matching on axial-flow-compressor performance. Trans. ASME J. Eng. Gas Turbines Power (August), 1273–1293.
Syverud, S., Bakken, L.E., Langnes, K., Bjornas, F., 2003. Gas turbine operation offshore—on line compressor wash at peak load. In: Proceedings of ASME Turbo Expo, New Orleans, June 4-6, 2003, ASME Paper No. GT-2003-38071.
Tabakoff, W., 1988. Causes for turbomachinery performance deterioration. In: ASME Gas Turbine and Aeroengine Congress, Amsterdam, June 6-9, 1988. ASME Paper No. 88-GT-294.
Tarabrin, A.P., Bodrov, A.I., Schurovsky, V.A., Stalder, J.P., 1998. Influence of axial compressor fouling on gas turbine unit performance based on different schemes and with different initial parameters. In: ASME International Gas Turbine and Aeroengine Congress, Stockholm Sweden, June 2-5, 1998, ASME Paper No. 98-GT-416.
Tarabrin, A.P., Bodrov, A.I., Schurovsky, V.A., Stalder, J.P., 1996. An analysis of axial compressor fouling and a cleaning method of their blading. In: ASME International Gas Turbine and Aeroengine Congress, Birmingham, UK, June 10-13, 1996, ASME Paper No. 96-GT-363.
Tatge, R.B., Gordon, C.R., Conkey, R.S., 1980. Gas turbine inlet filtration in marine environments. In: ASME Paper No. 80-GT-174.
Zaba, T., Lombardi, P., 1984. Experience in the operation of air filters in gas turbine installations. In: ASME International Gas Turbine and Aeroengine Congress, June 1984 ASME Paper No. 84-GT-39.
Zaba, T., 1980. Losses in gas turbines due to deposits on the blading. Brown Boveri Rev. 12–80, 715–722.

Bibliography

Bakken, L.E., Skorping, R., 1996. Optimum operation and maintenance of gas turbines offshore. In: International Gas Turbine and Aeroengine Congress, Birmingham, UK, June 10-13, 1996, ASME Paper No: 96-GT-273.
Becker, B., Bohn, D., 1984. Operating experience with compressors of large heavy duty gas turbines. In: International Gas Turbine and Aeroengine Congress, June 1984 ASME Paper No. 84-GT-133.
Benvenuti, E., Bettocchi, R., Cantore, G., Negri di Montenegro, G., Spina, P.R., 1993. Gas turbine cycle modeling oriented to component performance evaluation from limited design or test data. In: Proc. ASME IGTI Cogen Turbo Power, IGTI. vol. 8. pp. 327–337.
Bowen, T.L., Guimond, D.P., Muench, R.K., 1987. Experimental investigation of gas turbine recuperator fouling. ASME J. Eng. Gas Turbines Power 109 (July), 249–256.
Cohen, H., Rogers, G.F.C., Saravanamuttoo, H.I.H., 1987. Gas Turbine Theory. Longman, p. 1987.
Denton, J.D., 1993. Loss mechanisms in turbomachines. In: International Gas Turbine and Aeroengine Congress, Cincinnati, Ohio, May 24-27, 1993, ASME Paper No: 93-GT-435.
Diakunchak, I.S., 1991. Performance deterioration in industrial gas turbines. In: ASME International Gas Turbine and Aeroengine Conference, Orlando Florida, June 3-6, 1991, ASME Paper No: 91-GT-228.
Faddegon, C.J., 1999. Effective compressor cleaning as a result of scientific testing and field experience. In: Powergen Europe. Brussels. June.
Hoeft, R.F., 1993. Heavy-Duty Gas Turbine Operating and Maintenance Considerations. General Electric publication GER-3620B.

Lakshminarsimha, A.N., Saravanamuttoo, H.I.H., 1986. Prediction of fouled compressor performance using stage stacking techniques. In: 4th ASME Fluid Mechanics Conference on Turbomachinery Performance Deterioration, May 12-14.
Mathioudakis, K., Stamatis, A., Bonataki, E., 1999. Diagnosing the sources of overall performance deterioration in CCHT plants. In: International Gas Turbine and Aeroengine Congress, Indianapolis, Indiana June 7-10, 1993. ASME Paper No: 99-GT-364.
Upton, A.W.J., 1974. Axial flow compressor and turbine blade fouling, some causes, effects and cleaning methods. In: 1st Symposium on Gas Turbine Operation and Maintenance, National Research Council, Canada, October.

Index

Note: Page numbers followed by *f* indicate figures and *t* indicate tables.

FSC
www.fsc.org
MIX
Papier | Fördert
gute Waldnutzung
FSC® C083411

Kolibri 360 GmbH
Ferdinand-Jühlke-Straße 7
99095 Erfurt, Deutschland
produktsicherheit@kolibri360.de